Crack-Healing in Metals

David J. Fisher

Published by **Materials Research Forum LLC**
Millersville, PA 17551, USA

Published as part of the book series
Materials Research Foundations
Volume 181 (2025)
ISSN 2471-8890 (Print)
ISSN 2471-8904 (Online)

Print ISBN 978-1-64490-376-6
ePDF ISBN 978-1-64490-377-3

This book contains information obtained from authentic and highly regarded sources. Reasonable efforts have been made to publish reliable data and information, but the authors and publisher cannot assume responsibility for the validity of all materials or the consequences of their use. The authors and publishers have attempted to trace the copyright holders of all material reproduced in this publication and apologize to copyright holders if permission to publish in this form has not been obtained. If any copyright material has not been acknowledged, please write and let us know so we may rectify in any future reprint.

Distributed worldwide by

Materials Research Forum LLC
105 Springdale Lane
Millersville, PA 17551
USA
https://mrforum.com

Printed in the United States of America
10 9 8 7 6 5 4 3 2 1

Table of Contents

Introduction

The healing of cracks in metallic components has always been an attractive concept due to its economic and safety benefits. Traditionally, it was believed that cracking was irreversible. However, it has long been theorized that very small cracks could potentially be healed if the maximum temperature during heat-treatment allowed the atoms on either side of the crack to approach each other closely enough via thermal agitation. While improvements in crack-healing processes generally entail increased costs, they offer the advantage of producing low-strength materials more reliably and at lower cost if applied during an intermediate stage of the service life. From a physics perspective, crack healing fundamentally involves the exchange of material and energy within an open system. This means that cracks in metals can be healed by the effective importation of energy. The evolution of damage in metals suggests that various heat-treatment techniques may reduce the number of defects after prolonged exposure to high temperatures. Although many studies on crack-healing have focused on polymers, ceramics, and concrete—including capsule-based self-healing systems and shape-memory alloy-reinforced materials—it is more difficult to heal cracks in metals. This is primarily due to the high atomic bond strength and the low diffusion rates of atoms in metallic materials. Efforts have been made to incorporate shape-memory alloy wires into metal matrices to heal cracks via heating, and micro-cracks and voids in aluminum alloys have been significantly reduced through solid-solution treatment. While hot plastic deformation is another effective technique for repairing cracks, it cannot be applied to already finished parts. Various surface crack repair methods include plasma spraying, electroplating, and gas carbonitriding. Welding is a widely used approach, with friction stir welding capable of repairing cracks up to 2mm in width. Laser re-melting has also been used to repair fatigue cracks in steel by reducing both crack density and crack length. However, both welding and laser repair are unsuitable for internal cracks. Electric current-pulsing has gained popularity due to its applicability to nearly all metals and its relatively short processing time. Yet, its effectiveness is limited to simpler geometries such as plates, as complex shapes can lead to sparking. This method is more effective when combined with a subsequent heat treatment. The healing of internal cracks in damaged materials can be likened to second-phase sintering. Changes in plate-like defects resemble the breakdown of a second-phase plate via surface diffusion. This process occurs in two stages: in the first, atomic diffusion—driven by the chemical potential difference between the edges and the center—causes the rim of the defect to detach from its central part. In the second stage, Rayleigh instability leads to the fragmentation of the detached doughnut-shaped rim into a ring of spheres. A common method of healing is heat-treatment. For instance, cracks in α-iron can be healed by heating above 1000K. However, this technique is limited in effectiveness due to the low efficiency of self-diffusion mechanisms, restricting it to only very small cracks. Effective healing typically requires both high temperatures and long durations. Unfortunately, higher temperatures can adversely affect the microstructure and properties of the material. Nevertheless, specific methods such as copper precipitation have been successful in healing deformation-induced defects in under-aged Al-Cu-Mg alloys. This approach has also proven effective in repairing early-stage damage in austenitic stainless steels. The benefit of this mechanism lies in its ability to operate at lower temperatures, thereby avoiding the adverse effects associated with high-temperature exposure. However, only precipitation-strengthened alloys can benefit from this method. Electropulsing has been used to heal cracks in carbon and stainless steels. This technique relies largely on Joule heating generated by current accumulation at crack tips. However, the microstructure and mechanical properties of the melted and healed region remain unclear. Among all known methods, thermo-mechanical coupling is considered the most effective for healing large cracks, as it reduces crack width through the application of pressure. This method involves grain recrystallization and atomic migration in the affected region but is inapplicable if the workpiece cannot be further deformed. None of the aforementioned techniques—most of which rely on energy input—appear to be effective for healing large surface cracks. Due to the relatively low diffusion rate of metallic atoms, supplementing the material externally becomes more important in such cases. One commonly used method is the filling of cracks via oxide formation and the resulting volume expansion. However, this approach is not always effective for metals and alloys because the oxide film formed on the crack surface often inhibits further healing. Thus, it is believed that fast atomic diffusion and material filling without oxidation are critical for effective crack healing in metals. Thermochemical treatments are particularly suitable in this regard, as they create an ideal environment for crack healing. These treatments also help to prevent crack initiation by strengthening the material's surface. In additive manufacturing, printed micro-cracks can be healed through localized melting in the heat-affected zone and the introduction of compressive stresses. Gheysen et al. proposed a liquid-assisted hot isostatic pressing process for healing pores in Al-Mg alloys produced via laser powder bed fusion, which overcomes the traditional HIP limitation of being ineffective for aluminium alloys. The healing of metallic components is also attractive for its economic advantages, especially given the emergence of microcracks in structural metals. The process termed flash hot isostatic pressing, which integrates laser shock peening during laser powder-bed fusion, enables healing through localized melting and stress induction. The liquid-assisted hot isostatic pressing method also supports effective pore-healing in Al-Mg alloys prepared by laser powder-bed fusion. Crack initiation and propagation remain key causes of metallic component failure. Consequently, investigating methods for crack arrest and healing is a vital area of research in engineering and materials science. As noted, the primary energy-input methods currently used for internal crack healing include high-temperature heating, thermo-mechanical coupling and electric currents. For example, studies on low-carbon steel have shown that combining heating with external pressure accelerates crack healing between 900 and 1200C. Similarly, hot plastic deformation enhances healing through increased temperature and strain. However, these methods risk inducing grain-coarsening and producing large heat-affected zones due to extended high-temperature exposure. They also require complex heating and control systems and can introduce unwanted plastic deformation. To mitigate such limitations, electric current-based methods have been proposed. These include healing fatigue cracks in steel, repairing quenched cracks in 1045 steel, and micro-crack healing in titanium alloys using high-density electropulsing. Cracking is one of the most prevalent ultimate failure modes in materials. Healing or repairing cracks is desirable not only for cost-saving but also for ensuring long-term performance and safety. Generally, effective crack healing requires atomic mobility from a healing agent and the ability of those atoms to bond with the crack surfaces. Healing cracks in metals poses greater challenges than in polymers due to metals' high melting points and extremely low atomic mobility at room temperature. Hence, increasing the temperature is often necessary to enhance mobility. For example, micro-cracks in α-iron require heating above 1000K to be healed, although such high-temperature methods are effective only for very small cracks and can adversely alter material properties. At the nano-scale, open-volume defects such as vacancies and dislocations can be healed via precipitation. In Al-Cu-Mg alloys, solute copper atoms precipitate at defect sites and significantly reduce defect concentration, making this approach suitable for repairing early-stage damage. However, it is limited to alloys that undergo precipitation hardening. Electropulsing heals cracks by inducing local heating and compressive stresses in the crack region. Additionally, metal ions exhibit relatively high mobility in aqueous or molten-salt electrolytes, enabling electrocrystallization. In this process, a cracked metal part is used as a cathode, while metal ions in the electrolyte act as healing agents. Under appropriate electric potential, the metal ions are reduced to atoms, which then bond to the crack surfaces and heal the damage. Electro-healing involves mass transport, adsorption, diffusion, nucleation, and crystal growth within confined spaces. Its success depends on the electrolyte's throwing power and the applied current density. Self-healing alloys represent a promising category of materials that can mitigate the risk of failure. Crack formation can be detected using ultrasonic methods, and in many cases, immediate replacement of damaged parts is not possible. Therefore, the ability to heal cracks during operation—for instance, through annealing—is highly advantageous. It will be useful to survey the theoretical and practical application of the available crack-healing methods to metals.

Aluminium

Low-angle X-ray diffraction was used[1] to study the growth and healing of fine (100nm) cracks in aluminium which was deformed in tension and annealed. Micro-crack growth was observed mainly in sub-surface layers in the early stages of deformation. The thickness of the defect layer increased with increasing reduction. Healing of micro-cracks began immediately after heating, and involved vacancy-diffusion within a surface layer which was about 10μm thick. Healing in deeper layers required holding at temperature.

Molecular dynamics methods were used[2] to simulate micro-crack healing during heating or stressing. A centre micro-crack in an aluminium crystal could be sealed by imposing a critical compressive stress or by heating to above a critical temperature. During the micro-crack healing, dislocations and vacancies were generated and moved, with twins sometimes appearing. The critical temperature which was required for micro-crack healing depended upon the orientation of the crack plane; the critical temperature being lowest for cracks along the (111) plane. When pre-existing dislocations were present around the micro-crack, the critical temperature for micro-crack healing decreased. The energy condition for crack-healing was:

$$\sigma^2\pi\alpha(1-v^2)/E + \partial U/\partial A \geq 2\gamma + \gamma_P$$

where σ was the applied compressive stress, α was the crack-length, v was the Poisson ratio, E was the Young's modulus, U was the heat-energy driving the crack-healing, A was the area of the crack, γ was the surface energy and γ_P was the work of plastic deformation. The presence of pre-existing dislocations could reduce γ_P.

The passivation breakdown of aluminium-based alloys can be caused by mercury concentrations lower than 300ppm. When such alloys are used as sacrificial anodes, it causes a decrease of more than 0.3V in their operational potential in chlorides. Evidence was found[3] for the mechanism via which mercury produced its marked effect when present in non-aggressive aqueous media by performing experiments on 4N-purity aluminium in mercury acetate solutions of various concentrations and pH-levels. The results showed that, following immersion, an instantly-formed initial oxide film exhibited dynamic crack-healing at flaws in the film. These were possibly associated with grain boundaries. Any subsequent healing depended upon the composition of the medium. In this particular case, Hg^{2+} ions could be directly reduced on bare aluminium, initiating a surface diffusion which permitted the formation of an amalgam. Aluminium atoms then diffused through the liquid mercury and underwent oxidation at the amalgam/electrolyte interface. This led to oxide-detachment and to wide cavities. Aggressive anions were not required for activation initiation. The initiation of the activation process was likely to depend upon the medium's composition. Various Hg(II) acetate solutions, of differing concentration and pH were prepared. Aluminium specimens were placed in various solutions and their responses were monitored. An active state was deemed to be attained when the measured potential in sulphate solution was more negative than 1100mV. Upon comparing the activation times which were required for a given Hg^{2+} concentration, it was deduced that, at lower pH-values, healing at crack flaws was hindered. Changes in the Hg^{2+} concentration led to only slight differences in the activation time. It was more effective to hinder healing by using a low pH than it was to increase the cracking rate by using a stronger polarising couple, such as Hg/Hg^{2+}. Although a higher Hg^{2+} concentration increased the possibility of Hg^{2+} being to be reduced at the bare aluminium at a cracked flaw, only a small number of active sites was required to shift the corrosion potential to an extremely low value.

Molecular dynamics methods were used[4] to simulate the effects of ultrasonic cavitation on aluminium blocks which contained micro-cracks. This revealed that the micro-crack size tended to decrease or to close entirely following impact. After cavitation, numerous dislocations formed around the micro-crack tip, thus facilitating partial or complete closure via dislocation-shielding and atomic diffusion. The crack-healing process was closely related to the external forces which were generated by impact and by changes in surface energy within the crack. Following crack-healing, stresses in the matrix tended to congregate around healed micro-cracks, stacking faults and grain boundaries.

Pulsed electric current treatment can accurately locate cracks distributed within a material and permit *in situ* repair. The effect of the treatment upon a crack-containing sample was determined[5] by using finite-element methods. The electric current-density, temperature, stresses and displacements were concentrated in the crack area, and mutually interacted, thus driving crack-healing. Following pulsed electric current treatment, the mechanical properties of a crack-containing sample largely recovered and the tensile strength and elongation increased by 23.2% and a factor of about 2, respectively. Bonding and densification occurred between crack surfaces, and the healed interfaces had a higher strength. Recrystallization occurred in the crack area, large numbers of dislocations were continually emitted from the crack tip and grain boundaries progressively migrated toward the crack. Plastic deformation at high temperatures prompted the diffusion and migration between the two sides of the crack, and the formation of fibrous bridging structures. This provided new insights into the crack-healing mechanism of pulsed electric currents. Studying the evolution of microstructures during the crack-healing process clarified the crack-healing mechanism. Under the action of the current, crack-tips experienced compressive-stress concentration. The cause of this high compressive stress zone was two-fold. In agreement with finite-element method simulation results, the Joule-heating effect which was caused by the treatment caused the temperature to increase, and thermal expansion of the material produced compressive stresses that were directed towards the crack-tip. The electron-wind also generated compressive stresses at the crack-tip. The high compressive stresses caused the crack-tip to undergo a plastic deformation which made the material squeeze towards the middle, and encouraged atoms on both sides of the crack to move towards the middle and produced gradual healing of the crack from the tip. The crack-healed zone retained a distribution of compressive stresses even when the crack had disappeared. Following recovery to the room-temperature equilibrium state, the entire simulation is essentially a distribution of compressive stresses, with the compressive stresses in the crack-healed zone being higher. The continued existence of compressive stresses inhibited the re-initiation and propagation of cracks, further improving the healing effect. The healing of a crack could also be clearly described in terms of atomic displacements. Starting from the crack-tip, atoms on both sides of the crack underwent obvious opposed movements which were driven by the high compressive stresses produced by the treatment. Directional migration of atoms directly promoted bonding between the crack surfaces, making the atoms on both sides bond. A new crack-tip was thus formed below this point, and the distance between the crack surfaces below it was reduced, thus benefiting further healing of the crack. This process was then repeated continuously and the crack was gradually healed from top to bottom; the visual appearance being that the crack-length continually shortened until it disappeared. Along this direction of gradual healing of the crack, the displacements of atoms on both sides of the crack gradually increased: the atoms in the healed zone exhibited the largest displacements, and the overall number of displaced atoms also gradually increased. The region in which atoms underwent displacements was a superposed state of triangles and parallelograms; traces produced by dislocation proliferation and slip within the material. When the model simulation recovered to the room-temperature equilibrium state, atoms in the healed zone did not undergo reverse motions, thus indicating that crack-healing was stable and irreversible. Mixing occurred between the atoms on both sides, leading to metallurgical bonding between the crack surfaces, and effectively filling the crack. When high compressive stresses were driving crack-healing, the material underwent a clear plastic deformation which was closely related to the crack-healing mechanism. Microstructures such as dislocations and stacking faults were identified and visualized. During crack-healing, dislocations and stacking faults were continuously generated, accompanied by a downward movement of the crack tip, and accumulated in great quantities in the crack-healed zone. Most of the dislocations generated were Shockley partials, 1/6<112>, while the stacking faults were formed by slip of the Shockley partials on {111} slip-planes. When the model had recovered to the room-temperature equilibrium state, the dislocations were clearly reduced, especially in the healed zone while a large number of stacking faults remained. The plastic deformation behaviour during crack-healing was a clear reflection of the dislocation evolution behaviour within the material. Obvious dislocation-emission occurred at the crack-tip under the action of the current, and was a result of compressive stresses. Joule-heating also made some contribution to dislocation emission. The high temperature which was induced by it increased the thermal activation energy of atoms, and made it easier for atoms to overcome the energy barriers which were required for dislocation multiplication and motion. The high temperature also increased the activity and kinetic energy of atoms, and accelerated their movements. This accelerated the diffusion of atoms, thus contributing to the formation of dislocation nuclei and dislocation motion. During subsequent healing, large numbers of new dislocations were continually emitted from the new crack-tips, and slipped on {111} slip-planes, promoting diffusion and migration between the crack-surface atoms. Areas on both sides of the crack then merged, and an interwoven dislocation network formed in the healed zone. The emission and motion of a large number of dislocations promoted plastic deformation at the crack-tips, leading to blunting of the crack-tips and to a reduction in crack size, thereby increasing the fracture toughness of the material. During recovery to the room-temperature equilibrium state, the model always tended to minimize its total energy. This led to a gradual decrease in the speed of atoms, which also tended to migrate to stable lattice structures under the action of stresses. Local rearrangement of atoms eliminated some dislocations, and also prompted dislocations to move towards lower-energy forms. Due to the higher dislocation density in the healed zone, the dislocations underwent strong interactions during movement and resulted in dislocation annihilation and contraction. The numbers of dislocations in the healed zone of the model were thereby finally reduced and the dislocation-density decreased. When a crack-containing sample was subjected to the pulsed-current treatment, some of the current was forced to detour around, and concentrate, at the crack tip. This led to a sharp increase in the local current-density and produced more intense Joule-heating in this area. The high temperature caused the material to undergo recrystallization. Fibrous bridging structures, a key discovery here, joined the material on the two sides of the crack.

Al-Ag

Self-healing is a very desirable characteristic that can greatly improve the performance of aluminium-based alloys which are noted for a low density and good mechanical properties; especially when produced via rapid solidification. Self-healing can close cracks and, in the case of aluminium-based alloys, self-healing typically involves surface modification. Transmission electron microscopy was used[6] to observe the crack self-healing behaviour following annealing. Cracks were monitored in rapidly solidified Al-30Ag alloy which had a non-equilibrium phase composition which was made up of a small fraction of Ag_2Al and a supersaturated solid solution of silver in a face-centred cubic aluminium matrix. Following annealing (450C), an equilibrium phase composition was obtained by forming a larger amount of Ag_2Al. This phase transformation did not permit the crack to heal. Further annealing (550C) caused recrystallization into a supersaturated solid solution of silver in face-centred aluminium, followed by a return to a mixture of face-centred cubic aluminium and Ag_2Al upon cooling. This process was accompanied by closure of the crack. This confirmed the self-healing possibilities of a Ag_2Al phase, and this could be exploited by dispersing fine Ag_2Al particles in a structural material so as to endow the material with self-healing properties. The Ag_2Al had an hexagonal close-packed structure and this transformation did not involve a volume-change. This was an advantage with respect to reversible self-healing because volumetric changes would encourage crack-propagation. The above type of metal-matrix composite could be fabricated by using powder-metallurgy techniques such as hot-extrusion, spark plasma sintering, additive manufacturing and cold pressure-welding. Although the Al-30Ag alloy was not strong enough for direct structural use, it was useful as a source of fine dispersed Ag_2Al precipitates.

Al-Cu

Hot-cracking in Al-Cu direct-chill cast billets was investigated[7] by controlled casting. No hot-cracking was found in billets of alloys which contained more than 4%Cu. Decreasing the copper concentration from 3 to 1% led to a decrease in the amount of eutectic which was visible on hot-tearing surfaces. Special features appeared on the hot-tearing surfaces of Al-1%Cu alloy, where hot-tearing propagated mainly through the solid phase. A solute-rich (eutectic) path along grain boundaries, ahead of the hot-crack tip, was observed in billets with hot tears and was attributed to crack-healing. Porosity concentrated at the centre of billets, and its volume was lower in alloys which exhibited a high hot-tearing susceptibility, and in alloys which were not susceptible to hot-tearing. Numerous studies had demonstrated that binary alloys could have a specific compositional range within which the hot-tearing susceptibility reached a peak. Some work had extended such observations to binary aluminium alloys when processed using direct-chill casting. It was confirmed that increasing the content of an alloying element to beyond the peak susceptibility range reduced the occurrence of hot-cracking. In the aluminium-copper system, the maximum hot-tearing susceptibility occurred at about

1%Cu. Alloys which contained more than 4%Cu usually exhibited no sign of cracking during casting, and this behaviour was attributed mainly to the more favourable combination of a narrowed solidification range, which led to reduced thermal contraction, and to an increased amount of eutectic liquid, which increased feeding during solidification at high solid fractions. In billets with 2 to 3%Cu, hot-tearing fracture surfaces had an appearance which was consistent with previous observations and included dendritic patterns, with a liquid film covering the dendrites. At 2%Cu, solid bridges were present on the fracture surface. Hot-tearing propagated through a continuous liquid film or through a combination of liquid films and solid bridges. These distinct modes led to differing surface morphologies. The former mode produced a smoother fracture surface, while the latter mode produced a more irregular surface, due to broken solid bridges. The degree of grain-boundary bridging during solidification depended upon the alloy composition, with highly-alloyed materials had fewer bridges. This was related to the amount of eutectic which was present in the as-cast microstructure, because the non-equilibrium eutectic liquid was the final stage of solidification. The volume of eutectic increased with increasing copper content. A greater eutectic content increased the possibility of shrinkage compensation via liquid feeding and could aid crack-healing. A solute-enriched eutectic path extended over several grain boundaries and appeared to continue from a hot crack, suggesting that fluid movement occurred within the dendritic network and promoted healing. Eutectic-liquid feeding limited the formation of the shrinkage-cavities caused by solidification, thermal contraction and gas-evolution. When the eutectic content was low and not sufficient to form continuous liquid films, the grains consolidated via solid bridges and forced cracks to propagate through those connections. As the material solidified, it acquired strength due to grain-bridging and coalescence. The solidifying mushy material becomes able to transfer stress, thus permitting thermal-contraction stresses from the billet's outer shell to generate tensile forces at the centre. If the rate of strength-development exceeded that of stress-accumulation, the mushy material could resist deformation at higher solid fraction. This resistance is effective only in the absence of stress-concentrators such as pores, residual liquid films or segregated liquid.

Al-Mg

Crack-healing by using electric currents involves complex interactions between electrical, thermal, and mechanical fields. Crack-arrest using Joule heating has previously been simulated by using thermo-electro and structural coupling theory, and finite-element models were later used to analyse the melting of crack-tips in conductive plates. Other studies have investigated time-dependent temperature changes, temperature gradients and Von Mises stresses in metal plates during electropulsing by using finite-element methods.

Theoretical studies of coupled fields at the crack-tip provide an understanding of the crack-healing mechanism and guide practical solutions. One approach involves a determination of the electric field at the crack-tip by using complex-function theory and electrodynamic arguments. This reveals the local heat-source, and thus the corresponding temperature distribution. That method has been applied to edge-cracks, embedded elliptical cracks, co-linear cracks and half-embedded cracks. The relationship between electrical parameters and the driving force for crack-healing requires further quantitative analysis.

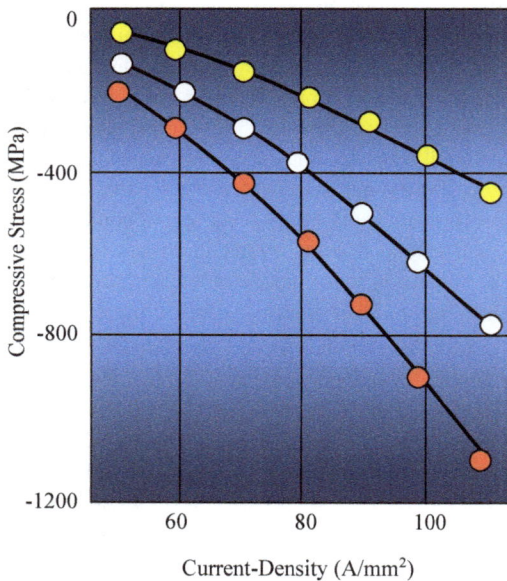

Figure 1. Temperature at 0.5mm from a crack tip in 2mm-thick Al-2.2Mg-0.25wt%Cr alloy as a function of current density and crack-length. Red: a = 6mm, white: a = 4mm, yellow: a = 2mm

That was done here by applying complex-function theory to the coupled fields near to the crack-tip in a metal sheet having an edge-crack which was subjected to an electric current. The energy release-rate was used to evaluate the driving force for healing at various current densities. Experiments were performed[8] on 2mm-thick sheets of 5052-

H32 (Al-2.2Mg-0.25wt%Cr) having an electrical conductivity of 2 x $10^7/\Omega$m, a thermal conductivity of 138W/(mC), a Young's modulus of 70.3GPa, a thermal expansion coefficient of 2.25 x 10^{-5}/C and initial temperature of 20C. High temperatures at the crack tip, which were often above the melting point, explain tip-melting. The current density greatly affected the temperature and the thermal compressive stresses near to the crack tip. Both factors increased with current density and crack length, thus increasing atomic diffusion and crack-healing. The energy release-rate served as a measure of the driving force for healing. It could be calculated theoretically and by using finite-element simulations via the J-integral.

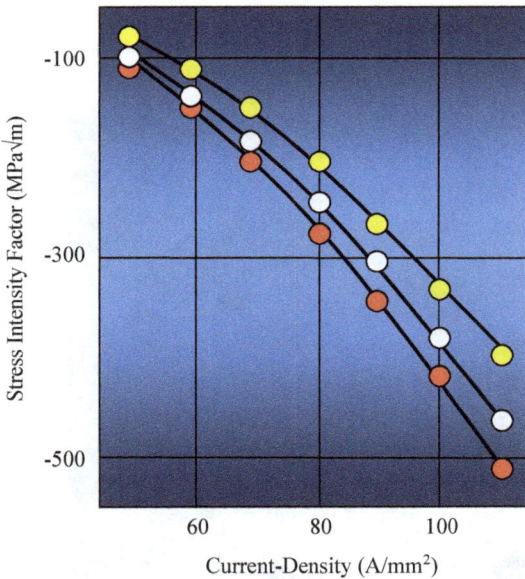

Figure 2. Thermal compressive stress at 0.5mm from a crack tip in 2mm-thick Al-2.2Mg-0.25wt%Cr alloy as a function of current density and crack-length. Red: a = 6mm, white: a = 4mm, yellow: a = 2mm

The simulation results closely matched theoretical predictions. The energy release-rate increased markedly with current density and confirmed the suggestion that a higher current-density increases healing. On the other hand, excessive Joule heating could damage the matrix by, for example, forming dimples around the crack tip. In order to optimize healing while maintaining structural integrity the current-density must be carefully chosen. The distribution of electric fields, temperature fields and thermal-stress fields near to crack tips (figures 1 to 3) was analysed theoretically for the case in which an electric current was applied to a metal plate containing an edge crack. The current-treatment of the 5052-H32 alloy plate containing a fatigue crack, investigated experimentally and by finite-element simulation, showed that the current-density and crack-length had marked effects upon the temperature, thermal compressive stress and thermal-stress intensity-factor near to the crack tip. Micro-welding occurred due to heat-concentration and thermal compressive stresses around the crack tip. Simulations showed that there was a high heat-flux at the crack-tip, due to the effect of current-flow. Annular thermal compressive stresses, caused by the temperature gradient, formed around the crack-tip. These prevented crack-propagation and aided crack-healing.

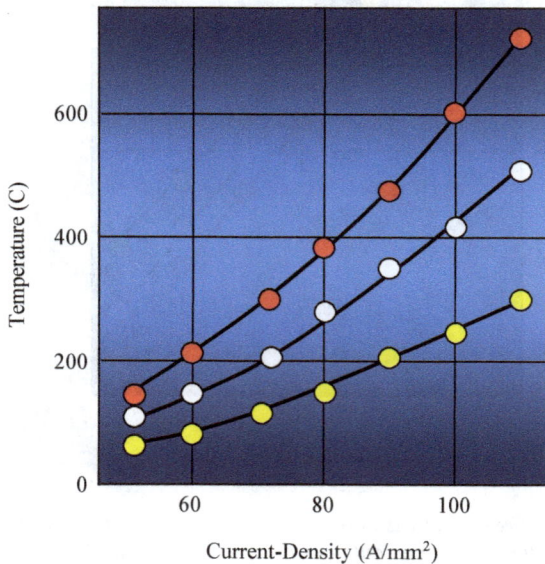

Figure 3. Thermal stress intensity factor at 0.5mm from a crack tip in 2mm-thick Al-2.2Mg-0.25wt%Cr alloy as a function of current density and crack-length. Red: a = 6mm, white: a = 4mm, yellow: a = 2mm

The effect of hot-rolling upon the growth of transverse cracks and surface oxides in aluminium-magnesium alloys was investigated[9]. With 4 passes, cracks were observed on the alloy surface following the initial pass. These cracks occurred along, and across, grain-boundaries and propagated into the sub-surface along grains. Nanocrystalline grains of MgO having an hexagonal structure were present on the crack faces, and sealed narrow regions of the cracks. Crack-healing began with the initial rolling pass, via the sealing of cracks with MgO. After 4 rolling passes, all of the cracks propagated along grain boundaries within the sub-surface regions. The hexagonal structure of MgO grains was no longer visible at this rolling stage, but all of the cracks were filled with porous MgO layers. By this rolling stage, the microstructures of the oxides within cracks and in the near-surface region were similar. Rolling stresses were deemed to contribute to the structure of the MgO within transverse micro-cracks during hot-rolling. Grain-sliding and step-formation led to overlapping at, and within, grain boundaries, and was accompanied by crack-formation. Nano-cracks appeared on the alloy surface. The micro-cracks which were observed at this stage extended into the sub-surface and were attributed to steps generated during rolling. The flattening-effect of further rolling removed grain steps and any overlapping at the grain boundaries. The diffusion of magnesium to free surfaces which were created by nano-cracks or micro-cracks promoted MgO formation and effectively filled the cracks. This was more marked in aluminium alloys having a high magnesium content. Wider cracks thus had MgO layers on their surfaces, while narrower cracks were sealed with MgO following the first rolling pass. Magnesium diffusion occurred near to intermetallic particles. Following the fourth rolling pass, micro-cracks propagated further into the sub-surface, but were shallower than those which formed after the first pass. The presence of the MgO within cracks prevented their closure during additional rolling, but the formation of MgO aided crack-healing by filling the gaps between crack faces: it sealed them and could block their deeper propagation. This effect was known from the oxidative healing which occurs in self-healing materials. During continued rolling at high temperatures, magnesium diffusion into the free surfaces of cracks began to fill the voids between crack-faces. The nanocrystalline nature of MgO gave the cracks a dark appearance, starting from the second pass, and was suggested to be caused by rolling-induced stresses. As a check, some samples were rolled just once and were then held at the second pass temperature of 525C for a time that corresponded to a rolling cycle. When their surfaces were compared with those of samples which were subjected to 2 rolling passes, the 2-pass samples had darkened grain boundaries while single-pass samples, which were exposed to heating alone, did not. This demonstrated the effect of mechanical stresses upon the development of MgO nanostructures. While MgO formation aided crack-healing by sealing them, it also inhibited satisfactorily complete welding of the crack faces.

Wrought aluminium alloys suffer from hot-cracking during selective laser melting. A modified AA6061 alloy was used[10] to obtain nearly-dense printed samples from powders which were mainly spherical and flowed smoothly. The results indicated that the addition of a small (0.15wt%) amount of scandium could heal cracks during selective laser melting. It could also markedly increase the strength of AA6061 to more than 350MPa, together with an elongation of 31%. A very fine microstructure within a melt-pool was created, with the formation of a nanometre to sub-micron bundle. Scandium and aluminium share a similar lattice structure, so that many aluminium atoms could be replaced with scandium atoms which acted as nucleation sites and prevented the formation of micro-cracks. X-ray diffraction studies were made of 4 types of sample: monolithic AA6061 powder, printed AA6061, AA6061-0.15Sc powder and printed AA6061-0.15Sc. The results revealed the presence of a primary α-aluminium phase plus a minor amount of Mg_2Si intermetallic. During conventional preparation, the Mg_2Si usually formed following solution-treatments and aging. In the present case, its presence was instead attributed to the thermal cycles which are involved in selective laser-melting. During this process, each previously-printed layer was subjected to rapid cooling (circa $10^7C/s$), followed by a re-heating which resulted from subsequent layer-deposition. The absence of an Al_3Sc phase indicated that scandium remains entirely in solid solution. Micro-cracks were observed, in monolithic AA6061 samples, which ran along both the build and transverse directions. No such defects were found in AA6061-0.15Sc samples, thus reflecting the effective crack-healing ability of scandium. The microstructure of the monolithic AA6061 resembled the dendritic pattern which is typical of that of cast AA6061. The AA6061-0.15Sc samples had a fine columnar microstructure within each melt-pool. Electron back-scatter diffraction revealed a refined structure which comprised tubular and knotted features. The tubes were less than 40μm in length, with diameters that ranged from the tens-of-nanometre to the sub-micrometre scale in going from the boundaries to the centre of the melt pool. This implied that the heating effect of the laser did not extend beyond individual melt pools.

The high-strength AA7XXX aluminium alloys are susceptible to hot-cracking during fusion welding. The liquation-cracking susceptibility of $T\text{-}Mg_{32}(AlZn)_{49}$-strengthened Al-Mg-Zn alloy with a Zn/Mg ratio of less than unity was studied[11] via circular-patch welding. The results were compared with the liquation-cracking tendency of $\eta\text{-}MgZn_2$-strengthened AA7XXX alloys in which the Zn/Mg ratios were greater than unity. All of the novel Al-Mg-Zn alloys exhibited as low a liquation-cracking susceptibility as did AA5XXX alloys, and greatly exceeded that of normal AA7XXX alloys. An increase in the Zn/Mg ratio resulted in a greater difference between the fraction of solid in the molten zone and the partially molten zone during final solidification. This could lead to a wider crack-healing difference between those two areas, and result in differing liquation-cracking tendencies in various alloys. In order to evaluate the liquation-cracking susceptibility of the various alloys, circular-patch welding tests were carried out. That is, each specimen was firmly clamped between 2 thick copper plates so as to prevent thermal contraction during welding and cooling.

Table 1. Percentage of micro-cracks
in aluminium-alloy welds

Alloy	Micro-Cracks(%)
1	0.16
2	0.40
3	0.85
4	1.52
5	5.66

This arrangement imposed symmetrical and consistent constraint-forces on the partially-melted zone and on the fusion zone. The frequency of micro-cracking increased with increasing Zn/Mg ratio, and indicated that higher Zn/Mg ratios increased the tendency to liquation-cracking. Alloy 5 (tables 1 and 2) exhibited a cracking-susceptibility which was 3 times higher than that of the other alloys. The cracking tendencies of Alloys 2, 3 and 4 were comparable to that of AA5083 alloy, confirming that the new Al-Mg-Zn alloys shared the low cracking-susceptibility of standard AA5xxx alloys. Non-equilibrium simulations of solidification behaviour were also carried out, showing that during final solidification, where the solid fraction was greater than 95%, the partially-melted zones always contained higher solid fractions than did fusion-zones in all of the alloys. This led to more liquid being in the fusion-zones and thus improving micro-crack healing in those regions. The increasing gap in solid fraction between the partially-melted and fusion zones, with increasing Zn/Mg ratio, explained the growing disparity in crack-healing efficiency and thereby the increased cracking-susceptibility. Alloy 1 exhibited the smallest difference in solid fraction between partially-melted and fusion zones, and this correlated with the lowest cracking-tendency as observed in AA5083. On the other hand, Alloy 5 had a much higher solid fraction in the partially-melted zone, as compared to that in the fusion-zone during the entire solidification process. This corresponded to the highest observed liquation-cracking tendency, which was typical of AA7039.

Table 2. Experimental aluminium-alloy compositions

Alloy	Mg(wt%)	Zn(wt%)	Zn/Mg	Mn(wt%)	Cu(wt%)	Cr(wt%)
1	5.0	0	0	0.80	0.15	0.03
2	5.0	1.0	0.2	0.80	0.15	0.03
3	5.0	2.0	0.4	0.80	0.15	0.03
4	5.0	3.0	0.6	0.80	0.15	0.03
5	3.0	4.0	1.3	0.25	0.05	0.2

Al-Si

The effects of adding 0.5, 1, 1.5, 2 or 3wt%Cu upon the quality-index and hot-tearing susceptibility of A356 (Al-7Si-0.35Mg) were investigated[12]. Additions of 1.5%Cu increased the quality-index by nearly 10%, and this was attributed to the solid-solution strengthening and dispersion-hardening effects of copper-rich Al_2Cu and AlMgCuSi. Further copper additions of up to 3% decreased the quality-index by nearly 12%. This was attributed to a reduction in the tensile strength and elongation which was caused by an increased volume fraction of brittle copper-rich intermetallic compounds and micro-porosity in the microstructure. Copper also increased the hot-tearing tendency of A356 alloy, as measured by using constrained rod-casting. Copper widened the solidification range of the alloy, thus decreasing its fluidity and increasing the period during which the alloy was in the zone of hot-tearing susceptibility. The hot-tearing fracture surfaces of high-copper alloys were rough, and the presence of interdendritic/intergranular micro-cracks was firm evidence for the initiation of hot-tearing in the later stages of solidification, where there was insufficient time for crack-healing. A bumpy appearance of some hot-tearing surfaces confirmed the presence of interdendritic liquid at the onset of hot-tearing, before the coalescence of dendrite-arms. Smooth fracture surfaces and a few intergranular tears also implied that sufficient amounts of eutectic liquid could flow into the interdendritic voids and cracks so as to heal them.

Aluminium-silicon alloys exhibit a lower hot-tearing tendency than do other aluminium alloys, but hot-tearing nevertheless sometimes occurs during casting. Additions of strontium reduced the hot-tearing tendency because of an increase in crack-healing. The effect of strontium upon the tensile strength, ductility and viscous behaviour during eutectic solidification was not clear however. A new investigation[13] of the effect of strontium additions involved tensile testing during the eutectic solidification of Al-2Si and Al-2Si-Sr alloys. This showed that the strontium additions did not affect the viscous properties, but the tensile strength and elongation during the later stage of eutectic solidification were increased by the additions. The hot-tearing reduction due to strontium additions was attributed to the improvement in ductility. It was suggested that an increase in the elongation due to the additions resulted from the decrease in tensile stress, given that this affected the initiation and propagation of hot-tearing due to an increase in porosity near to the crack.

Al_3Ti

Sintering methods can easily introduce micro-defects, such as centre-line faults, into intermetallics which then constitute weak points and act as fracture sources during deformation. Electropulse treatment (table 3) can typically heal cracks locally; i.e. at the narrow region in the middle of the crack. The wide regions at the tip of crack are not healed completely. The effect of electropulse treatments upon the microstructural evolution, mechanical properties and crack-healing of Al_3Ti was studied[14]. Compressive stress-strain curves revealed that the failure strain of Al_3Ti was increased by both electropulse and conventional heat treatments, but the increase was greater for the former. Fracture surface analysis showed that the intrinsic brittle fracture of Al_3Ti changed to quasi-cleavage fracture with increasing electropulse frequency. The effect of electropulse treatment upon Al_3Ti was related to an increase in atomic diffusion and dislocation movement, due to the coupling of thermal and athermal effects by the electropulse treatment. The thermal compressive stresses and athermal effects could also heal local micro-cracks and decrease the crack-width in Al_3Ti. Morphological changes in a micro-crack following electropulse treatment indicated that existing damage in Al_3Ti could be recovered, to some extent, by electropulsing. Due to the presence of cracks in the Al_3Ti, the resistivity distribution was inhomogeneous, and regions with a crack were usually higher in resistivity than were those without cracks. When high-energy pulses flowed through the alloy, the temperature increase caused by Joule heating of regions with cracks was higher than that for regions without cracks. This inhomogeneous temperature distribution could induce an inhomogeneous thermal expansion between the cracks and the matrix. The treatment produced an ultra-rapid temperature increase and thermal expansion, but the thermal expansion lagged behind the temperature increase. Regions with cracks therefore experienced thermal compressive stress, with a maximum value of $Ea\Delta T$, where E was the Young's modulus, a was the coefficient of thermal expansion and ΔT was the maximum temperature change. When ΔT was assumed to be 805C, E was taken to be 216GPa and a was 1.95 x 10^{-5}/C, the compressive stress was then about 3.44GPa. So the overall mechanism was that electropulse treatment could accelerate atomic diffusion and dislocation movement, thus facilitating atom and dislocation movement into regions with cracks. Under the influence of thermal compressive stresses, micro-cracks meanwhile healed locally and the crack-width decreased. Due to the coupling of thermal and athermal effects caused by pulsing, the grain-size of this alloy markedly increased in comparison to that produced by conventional treatments. The grain-size increased with increasing electropulse frequency. The athermal effect accelerated atomic diffusion and improved dislocation movement. When the electropulse frequency was increased to 350Hz and 450Hz, the intrinsic brittle fracture of this alloy became quasi-cleavage and the stress-strain curves had a relatively long non-linear stage. This indicated that the plastic deformation ability was improved by electropulsing. As compared with conventional treatments, the microhardness sharply decreased due to a softening effect of the treatment and its rate of decrease increased with increasing electropulse frequency.

Table 3. Parameters of conventional and electropulse treatments of Al$_3$Ti

Treatment	Frequency(Hz)	RMS Current-Density(A/mm²)	Temperature(C)
electropulse	250	7.9	432
electropulse	350	9.15	726
electropulse	450	10.3	885
conventional			432
conventional			726
conventional			885
crack-healing	450	10.2	805

Al-Zn

When Al-5.5Zn-2.4 Mg-1.3Cu alloy is cast into metal moulds, it has a high tendency to suffer hot-tearing, with a volumetric shrinkage due to density-differences during the transition from liquid to solid that generates shrinkage stresses. An absence of feeding of the remaining liquid during later solidification could also lead to hot-tearing. The relationship between solidification shrinkage displacement and stress was determined by using a device which was equipped with stress and displacement sensors[15]. This revealed that a reduction in grain-size changed the solidification shrinkage and reduced the solidification-shrinkage stress. This avoided the hot-tearing which was caused by high solidification-shrinkage stresses and local stress-inhomogeneity. Grain-size reduction also increased the number of feeding channels during solidification, thus making more liquid phase available for crack-healing.

Experimental data on 7050 aluminium alloy subjected to single-pass hot-compression were incorporated[16] into a cellular automaton model for microcrack-healing. The cellular automaton modelled the microstructural evolution of intergranular micro-cracks during thermoplastic healing, and took account of the topological deformation mechanism, dislocation-density evolution and dynamic recrystallization kinetics. Depending upon the crack surface and grain boundary, the crack surface-energy and grain-boundary energy drove grain-growth in various ways. The simulation results showed that hot thermoplastic deformation and dynamic recrystallization could entirely repair micro-cracks under some conditions. The appearance of crack-healing depended upon the crack morphology, the crack nucleation-rate and nucleation site, the growth direction and rate of new grain growth.

When metallic composite laminates are prepared by roll-bonding, a work-hardened surface layer is usually produced on the plate surface in order to promote bonding. Differences in the mechanical properties of the work-hardened layer and the substrate can however lead to the layer separating from the steel substrate during roll-bonding, with a resultant low bonding-strength of the laminate. A study was made[17] of steel/aluminium laminates with the aim of healing cracks between the work-hardened layer and the steel substrate by means of heat treatment. The aluminium component was AAA16061 and the steel was Q235 (0.16C, 0.4Mn, 0.3wt%Si). Following roll-bonding, the work-hardened layer of the steel surface had separated from the substrate, and the steel/aluminium laminates exhibited a lower bonding strength. Following annealing (450C,1h), laminates with cracks between the work-hardened layer and the steel substrate, a 426.8nm Fe_2O_3 layer was present between the work-hardened layer and the steel substrate; thus healing the cracks. Good connectivity existed between the hardened layer and the steel substrate.

The shear strength of the steel/aluminium laminates increased by 70%, following formation of the Fe_2O_3 layer. Shear fracture occurred mainly at the interface between the work-hardened surface layer and the steel substrate before the formation of the Fe_2O_3 layer. Following formation of the oxide layer, shear fracture occurred mainly in the aluminium. This approach therefore not only preserved the thin-film effect of the work-hardened surface layer, but also lessened the negative effect of delamination upon the overall bond-strength. Before roll-bonding, the prepared slab was heated to 450C at the rate of 45/min and held for 600s. Bonding was carried out by using a roller-diameter of 220mm at a rolling speed of 50mm/s, to a reduction of 45%. The laminates were then air-cooled to room temperature, followed by heat-treatment (450C) so as to induce healing. The suggested healing mechanism was based upon microstructural and compositional differences that arose during processing. Due to surface-brushing, variations in the mechanical properties and oxygen concentration arose between the work-hardened surface layer and the steel substrate. During rolling, the layer broke into smaller fragments and separated from the steel at points with high gradients of oxygen concentration. Aluminium from the substrate was extruded into the resultant cracks so as to form localized high-strength bonds with the steel substrate. These metallurgical bonding zones were concentrated around surface-layer fragments and remained in close contact with the layers, thus shortening the diffusion paths during subsequent annealing. During annealing, oxygen atoms from the surface layer diffused into the steel substrate and formed a solid solution with iron atoms which were near to the interface. The oxygen and iron later reacted so as to form Fe_2O_3 precipitates which bridged the gap between the surface layer and the substrate, thus greatly improving the bond strength. The crack-healing behaviour was very dependent upon the annealing temperature and duration. Increasing such factors can improve the healing of single-phase metals, but this steel/Al system was more sensitive. This was due to the possible formation of intermetallic phases at the interface, thus weakening the bonding strength. It was known that, when a work-hardened surface layer disintegrates, the resultant interface aids more rapid diffusion, thus increasing the possibility of intermetallic compound formation. Careful control of the annealing conditions was essential in order to heal the layer/steel interface completely, while avoiding the formation of intermetallic compounds. During heating at 500C for 1h, there was considerable intermetallic compound formation at interfaces; lifting the surface layer and breaking contact with the substrate. During heating at 450C for 1h, no intermetallics were seen and complete healing occurred, thus maximizing the bond strength. Annealing for longer periods again led to intermetallic compound formation, thus reducing the bond strength. Upon heating at 400C for 1.5h, partial healing occurred, without intermetallic formation, although some cracks remained and limited the strength.

Cobalt

Solidification-cracking during additive manufacturing impedes the use of that method. A $Co_{34}Cr_{32}Ni_{27}Al_4Ti_3$ high-entropy alloy which was susceptible to crack formation was prepared[18] by means of selective laser-melting. Many macroscopic cracks were present. Cracking was controlled by introducing iron-based metallic-glass powder as a glue during selective laser-melting. With increasing mass fraction of the metallic glass, the main defects which were present changed from cracks, to lack-of-fusion defects, and finally disappeared. The metallic glass preferred to segregate at the boundaries of the molten pool. Coarse columnar crystals gradually transformed into equiaxed crystals within the molten pool and into fine equiaxed crystals at the edge of the molten pool. This inhibited the nucleation of cracks and enhanced grain-boundary strengthening. Precipitates formed at the boundaries of cellular structures and greatly contributed to strengthening. The composite possessed a high ultimate tensile strength and elongation. The addition of iron-based metallic glass powder, as glue, into selective laser-melted high-entropy alloys is expected to be an attractive means for healing cracks and increasing the mechanical properties of additively manufactured components. It was recalled that the factors which affect the initiation and propagation of cracks can be divided into surface tension and liquid back-filling. In the case of surface tension, the appreciable cooling-rate could create a correspondingly high residual thermal stress. High surface tension often provokes crack initiation and propagation in a fabricated part. The importance of liquid back-filling lay in the ability of the liquid to fill dendrite channels. When the temperature and volume fraction of residual liquid decreased during solidification, volumetric solidification-shrinkage and thermal contraction in the boundaries of large columnar crystals produced cavities and hot-tearing cracks that could span the entire length of a columnar grain and propagate through further intergranular regions. Evolution of the structure is a significant factor in crack-healing. Due to constitutional undercooling and heterogeneous nucleation at the molten pool, columnar crystals that are sensitive to solidification-shrinkage can transform into small equiaxed crystals upon adding iron-based metallic glass powder. The cause of the undercooling and heterogeneous nucleation is solute redistribution in the melt which leads to the segregation of iron-based amorphous phases and changes the state of the melt.

Copper

Studies of crack-healing in pure copper and carbon steel showed[19] that an inner crack could be healed at high temperatures, suggesting that the healing of a long crack could depend upon a so-called separate healing mechanism involving the formation of a partition. The result was that a long crack transformed into a row of minute smooth cracks and round holes. It was noted that the diffusivity of iron to the crack-healing region was greater than that of other elements.

Molecular dynamics methods showed[20,21] that a centre micro-crack in a copper crystal could be healed by a compressive stress or by heating, with the effect of these factors being additive. Dislocation generation and motion occurred during micro-crack healing. If pre-existent dislocations were present around the micro-crack, the critical temperature or compressive stress which was required for healing decreased. The higher the number of dislocations, the lower was the critical temperature or compressive stress. The critical temperature which was required for microcrack-healing depended upon the orientation of the crack plane. The critical temperature was lowest (770K) for a crack located on the (001) plane. The crack plane was inclined at 70° to the (111) slip-plane. When a central crack was formed, the structure was then relaxed at 40K until equilibrium was reached. Simulations were performed at 600, 650, 700, 750 and 800K, using a time-step of 7.5 x 10^{-4}ps. At 600 to 700K, a central crack remained open after 50ps of relaxation. At 750K, the crack began to heal after 3ps, accompanied by dislocation-formation. After 7.5ps, the crack had almost healed, although a void and dislocations remained. The void persisted after 50ps. At 800K, the crack completely closed after 9ps, leaving dislocations but no voids. At 40K, a compressive stress which was perpendicular to the crack-plane was applied at a loading rate of 0.13GPa/ps. At a stress of about 7.26GPa, partial crack-healing occurred, leaving a void and dislocations. When the stress was increased to 13.2GPa, the crack was fully closed, with dislocations remained. The simulations indicated that the critical compressive stress and temperature for complete healing were about 13.2GPa and 800K, but when a compressive stress of 3.96GPa was applied at 600K the crack was closed completely after 11.25ps, leaving some dislocations. This showed that a compressive stress reduced the temperature which was required for healing. At 5.28GPa, healing occurred at 500K. Compressive stress and heating exerted synergistic effect upon crack-healing (figure 4). The crack-plane orientation markedly affected the critical healing temperature. That is, along the (111) plane, healing required 850K. Along the (001) plane, healing occurred at 770K. Along the ($\bar{1}$10) plane, healing occurred at 790K. In each case, healing was associated with dislocation generation and movement. A tensile stress was applied, using a loading-rate of 0.079GPa/ps, to an edge-cracked crystal. At 3.95GPa, a first Shockley dislocation was emitted. At 4.74GPa ($K_I \approx$

$0.6\text{MPam}^{1/2}$), 2 dislocations were released. These configurations were introduced into a centre-cracked crystal. For a centre-cracked crystal without dislocations, the critical healing temperature was 800K. When 2 Shockley dislocations were pre-inserted, healing began at 650K (2500 steps) and the crack was fully closed at 10000 steps, with new dislocations forming. The addition of 4 pre-existing dislocations reduced the healing temperature to 500K, with complete closure occurring after 6500 steps. At 40K, a compressive stress of 9.9GPa was sufficient for healing with 2 pre-inserted dislocations; reduction from the 13.2GPa which was needed without them. This confirmed that pre-existing dislocations lowered the critical temperature and the stress which were required for crack-healing. They were hypothesized to do this by reducing the resistance to atomic motion at the crack-tip. The elastic energy arising from compressive stresses and the thermal energy arising heating were additive driving forces for healing.

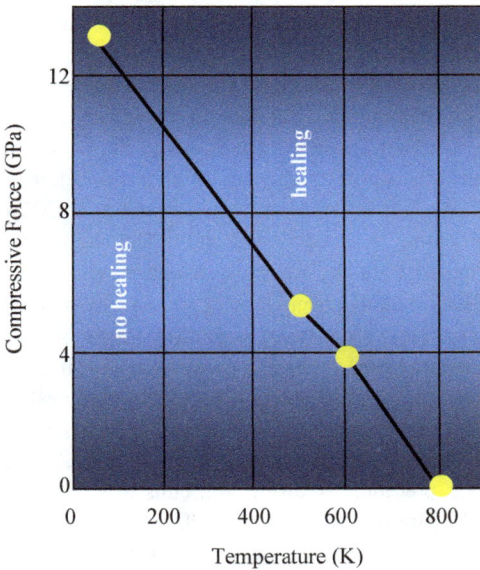

Figure 4. Region of crack-healing of copper as a function of temperature and compressive stress

In situ observations were made[22] of internal crack-healing in sections of pure copper by using high-temperature microscopy. A fine structure of healed micro-holes was observed by means of scanning electron microscopy. This showed that micro-voids with sizes of tens of microns began to heal at 750C, and were clearly healed at 900C. The fine structure of the healed micro-holes consisted of agglomerated particles at the micrometre-scale, plus much smaller particles on the agglomerated particles. The healing process involved passivation of the hole edge, shape-regularization of the micro-hole and inner interface motion of the micro-hole in the matrix.

Crack-healing is a thermodynamically irreversible process. A dissipation equation for healing was based[23] upon the thermodynamics of continuous media and the second law of thermodynamics. A variable was used to describe the crack-healing process, that was based upon molecular dynamics simulations. Molecular dynamics methods were used to simulate the healing of an ellipsoidal crack within a single copper crystal under a compressive stress. This showed[24] that dislocations were emitted from the ellipsoidal crack and moved on the $(1\bar{1}1)$ and $(\bar{1}11)$ planes under a constant compressive stress of 342MPa. The ellipsoidal crack shrank until it healed via dislocation emission, and annihilation at surfaces. Following crack-healing, residual dislocation nets and vacancy sites remained. The space within the ellipsoidal crack was, in effect, transferred to surfaces via dislocation emission and annihilation while the crystal underwent plastic deformation and the surfaces were roughened by the dislocation annihilation. The simulation showed that, when the temperature was greater than 200K, the energy-distribution of the lattice atoms widened significantly. The site of the dislocation core could then not be determined from the energy distribution, and it was not possible to simulate 3-dimensional crack-healing during heating. In the case of a penetrating crack in quasi 3-dimensional simulations which used periodic boundary conditions, all of the dislocations which were emitted from the crack-tip were straight lines that penetrated the thickness. Quasi 3-dimensional simulations indicated that crack-healing and dislocation emission and motion during heating were the same as that under compressive stress or under a combination of compressive stress and heating. It was proposed that crack-healing during heating was preceded by dislocation emission and motion, just as under compressive stresses. In the case of a penetrating crack in a small crystal which contained some 6000 atoms, the critical compressive stress which was required to induce crack-healing was, for example, 6.5GPa for aluminium and 7.26GPa for copper. The critical compressive stress for healing of an ellipsoidal crack in a crystal which contained 400000 atoms was 0.34GPa. This was an order-of-magnitude smaller than that for a penetrating crack. It was possible that an ellipsoidal crack could be more easily healed than could a penetrating crack. The healing of disc-like cracks in polycrystalline copper under uniaxial compression at 600C was studied experimentally[25]. A theory of crack-healing which was based upon a diffusion-dislocation mechanism was developed. A comparison of the experimental data with theoretical predictions revealed a close correspondence which was supported by an agreement of the experimental Peierls threshold stress with published data.

Three-dimensional molecular dynamics simulations were used[26] to determine the mechanism of room-temperature crack-healing in terms of dislocation-shielding and atomic diffusion with regard to crack-closure in a copper sample under shear loading. This showed that crack-healing occurred via dislocation emission at a crack-tip, together with intrinsic stacking-fault ribbon formation in the crack-tip wake. Dislocation-slip occurred in the matrix, and dislocation-annihilation occurred at the free surface. A crack closed rapidly, with the assistance of atomic diffusion induced by thermal activation, when the crack opening displacement was less than a threshold value. The crack-healing depended mainly upon the crystallographic orientation of the crack and upon the direction of the external loading. On the basis of energy characteristics and the crack-size effect, a theoretical model could predict the relationship between crack-size and shear stress. The molecular dynamics simulations revealed, in detail, that dislocation-shielding around the crack-tip was an essential feature of crack-healing. This effect introduces compressive stresses, and aided crack-closure via atomic diffusion. Simulations were made of a copper plate, containing a horizontal crack, that was subjected to an external shear stress. The crack inertially underwent elastic deformation. Dislocation emission began at this stage, and the crack exhibited partial closure at the tip, with this behavior resulting from appreciable plastic deformation of both crack surfaces; driven by dislocation rearrangement and internal slip. In order to accommodate shear strains, dislocation-slip occurred at the crack tip. Deformation-twinning was meanwhile initiated at the plate's upper and lower surfaces, due to high local stresses. Leading partial dislocations evolved into full dislocations, while micro-twins formed and contributed to further plasticity. Continued shear increased dislocation mobility, reduced the activation energy for diffusion and promoted rapid crack-surface closure near to the tip via thermally-activated diffusion. Dislocation-nucleation around the crack-tip generated local stress-fields which reduced crack-opening. As the dislocations moved, the stress-concentration increased and thereby accelerated closure. The crack-tip blunting was controlled by a competition between dislocation-emission and twinning. This decided whether fracture was ductile, due to dislocation slip, was brittle, due to twinning. Crack-closure occurred as dislocation activity overcame surface tension, and diffusion barriers. At an average shear strain of 4.15%, Shockley partial dislocations nucleated near to the crack tip. An investigation was also made of cracks which were oriented at various angles. The healing mechanisms of the inclined cracks mirrored those of horizontal cracks. Dislocation nucleation and emission occurred at the crack tip, and twinning dislocations glided in the [111] direction. Due to the (111) plane alignment, intrinsic stacking-fault bands then intersected at an angle of about 90°. As the inclination angle was increased from 0° to 90°, the shear strain decreased from 4.15% at 0° to 2.91% at 45°. It then increased to 14.51% at 90°. Healing was more likely at 45°, where the shear-stress components could exert the maximum compressive forces on the crack surfaces. It was suggested that the 45° inclination was the optimum for crack closure. The dislocation-shielding played a more predominant role than did direct compression in promoting closure. Healing was also typified by the formation of intrinsic stacking faults and twinning, in agreement with other observations. In order to relate the atomic-scale behaviour to energetic changes, the system's potential energy was calculated. The latter exhibited a marked angle-dependence. Higher strains required more energy for plastic deformation and crack closure, and the energy-versus-strain slope increased with angle, due to a greater dislocation activity near to the crack tip.

Other data indicated that, whereas intrinsic stacking faults were usually angle-independent, other dislocation behaviours depended strongly upon the inclination angle. This caused inhomogeneous atomic distributions to exist near to the crack-tip. Inelastic atomic displacements involved minimal movement at the crack-front, and were greatest near to the top and bottom surfaces. This suggested that atoms migrated along the shear direction in order to heal the crack. Shear-stresses during healing confirmed that plastic deformation occurred near to the crack-tip, due to high compressive stresses and dislocation-nucleation.

A simplified theoretical model envisaged a copper plate having a flat crack of length, c, subjected to a shear stress, τ, under the assumption of elastic isotropy. Partial dislocations which were emitted from the crack-tip generated intrinsic stacking-faults, leading to crack-closure via dislocation interaction. A criterion for healing was established which was based upon the energy balance:

$$\Delta W = W + U_T - 2c\gamma_c \geq 0$$

where W was the elastic energy arising from the shear stress, U_T was the thermal energy, and $\gamma_c = -\gamma_o + \gamma_p/2$, where γ_o was the crack-surface energy and γ_p was the work of plastic deformation work. It is to be noted that, in this context, the surface energy promoted rather than hindered healing. Shorter cracks healed more easily, due to the lower energy demands. The crack-healing below a 6GPa shear-stress was incomplete because insufficient energy was available. Atomic diffusion and thermal energy reduced the energy-barrier and promoted healing.

Crack-healing was also experimentally studied in pure copper at 873K, with a uniaxial loading acting perpendicularly to the crack plane. Analysis of the data by using a kinetic equation for crack-healing that was based upon a dislocation-diffusion mechanism showed that, under these conditions, healing was controlled by the diffusive-dissolution of the dislocation prismatic loops which were generated. This was due to a vacancy evaporation, which was caused by the loop-curvature, and to the absorption of interstitial atoms which were generated at the crossing-points of dislocations. Both processes here greatly accelerated crack-healing, and included the migration of interstitial atoms in crowdion configurations.

Cu-Zr-Al

Tapered $Cu_{50}Zr_{42}Al_8$ alloy was prepared[27] by means of copper-mould suction-casting. Cubic $AlCu_2Zr$ and B2-ZrCu martensitic B19'-ZrCu phases formed along the radial direction on the metallic-glass matrix. The microhardness of the monolithic bulk metallic glass revealed the presence of a soft surface and a harder centre. Larger composites had a soft centre and a harder surface, due to the secondary phases. The amorphous matrix was toughened by a transition-induced plasticity effect of the shape-memory phase, and weakened and embrittled by the $AlCu^2Zr$ crystals. Following annealing, self-healing of the Vickers indentation cracks occurred due to thermoelastic transformation of the shape-memory crystals. During loading, stress-induced martensites transformed from B2 to B19', with an associated volume expansion. The reverse transformation changed B19' to B2, with an associated volume shrinkage. A restoring force which arose from structural stresses drove crack closure.

Gallium

Battery anode materials can suffer from a short cycle life, due to the large volume changes and non-healable fracture which occur during electrochemical cycling. Gallium thin-film electrodes were prepared[28] on stainless-steel substrates in order to study the self-healing behaviour of gallium anodes. Following 25 cycles, the characteristic size of the self-healing area had been reduced to 34μm and gradually decreased with increasing cycles at a temperature above the melting point. There was a large amount of fluorine, oxygen and carbon at the surface, which were solid-electrolyte components. The presence of this layer impaired the self-healing capability of gallium thin films, because the layer could attach to the crack surfaces. Gallium-powder electrodes ($4.3mg/cm^2$) were prepared by using a simple liquid-dispersion method. The size of the powder, 3.43μm, was smaller than that of the effective self-healing area. Electrochemical measurements indicated an improved durability of the powder electrodes, as compared with that of gallium thin-film electrodes. Following 25 cycles, the average crack size of the powder electrodes was 1µm, thus showing that the self-healing ability of gallium in a liquid electrolyte is limited.

Gold

Using surface-force apparatus, optical interferometry, microscopy and X-ray diffraction, a study was made[29] of how gold and platinum films sintered or cold-welded at the nanoscale so as to form continuous bulk films when initially rough surfaces having nanometre-scale asperities were pressed together. Coalescence of the ductile materials occurred suddenly, in the manner of a first-order phase transition, when a critical local pressure or interparticle separation was attained. Thermodynamic and kinetic features suggested that this was a common behaviour for ductile materials which interacted at the nanoscale. The films ranged from 5 to 100nm and were deposited onto 2µm-thick molecularly smooth mica substrates. A gold film which was trapped between two mica surfaces possessed all of the characteristics of the bulk material. The film was of uniform thickness over the entire macroscopic contact area. The thickness was roughly twice the mean thickness (10.0nm) of the layers originally deposited onto the mica surfaces. Slightly thicker coalesced film was explained by the trapping of highly-compressed sub-microscopic air bubbles within the so-called cold-welded film. The results indicated that films coalesced rather than simply coming into adhesive contact. The latter was distinguished from coalescence by arguing that, in adhesive contact, the original surfaces remained distinguishable from the bulk at the contact interface. Coalescence implied complete disappearance of the original surfaces or any memory of the initial contact interface. Subsequent separation-failure could occur anywhere within the material. Quantitative insight into the micro- and nano-scale so-called contact-mechanics and adhesion-mechanics" of the systems was obtained by performing loading-unloading cycles on each sample when the surfaces were in contact. A sudden drop in the contact-radius just before rapid separation indicated a discontinuous reduction in the effective surface or adhesion energy. This could be understood by considering the boundary to be a crack-tip which moved from the high-energy gold/gold interface to the lower-energy gold/mica interface. The surface-energy of gold was 1.14 to 1.41J/m², while that for the mica/gold interface was estimated to range from 0.18 to 0.29J/m². Separation occurred in two stages. Crack-like failure of the gold film was followed by the peeling of the gold from the mica, which had a much lower surface energy, until the surfaces finally detached. The dominant stage depended upon the unloading or separation-rate. At high separation/unloading rates, rupture occurred at the gold/gold boundary. The contact-radius at pull-off was high and the adhesion-force was governed by the higher gold-gold

Materials Research Forum LLC
https://doi.org/10.21741/9781644903773

adhesion-energy. At lower separation-rates the gold/gold boundary broke more easily, and early in the unloading cycle, after which separation continued via peeling of the gold from the mica. It was concluded that the results provided a deeper understanding of the mechanisms of coalescence, cold-welding, sintering and crack-healing.

Nanoporous gold having a low relative density was produced[30] by de-alloying dilute precursor alloys, Au_xAg_{1-x} alloys (x = 0.10, 0.13, 0.15 or 0.17). Drying of this material in air led to a volume contraction of 22 to 38%, and a ductile-brittle transition occurred during drying: Wet samples of as-received de-alloyed nanoporous gold, when tested in water, exhibited a tensile strain of 8 to 10% before fracture. Following drying, the fracture-strain of the same material fell to about 1% in tension. The ductile-brittle transition was not caused by the familiar formation of cracks, but was instead connected with the self-healing of micro-cracks during drying. The plasticity of the as-received de-alloyed nanoporous gold arose from the presence of high-densities of native micro-cracks. Bridging or deflection of these micro-cracks in tension led to a diffuse failure which explained the high irreversible tensile strain of the wet de-alloyed material. The native micro-cracks could heal during drying, via the cold-welding of nano-ligaments under capillary contraction forces. The healing of these micro-cracks suppressed the diffuse failure which was caused by crack-bridging or deflection and led to brittle fracture of the dried material in tension. Micro-cracks in the x = 0.15 material become fewer, or vanished, after drying. This suggested that the micro-cracks had healed during drying. Water was removed from de-alloyed x = 0.15 material by freeze-drying. Most native cracks were retained in the freeze-dried material. The volume-contraction (7%) during freeze-drying was much lower than that (22%) produced by air-drying. The contraction in the former case was slightly non-uniform and led to the creation of additional large cracks. Re-immersion in water could introduce yet more cracks. Many of the cracks, especially the large ones, could not be fully healed by air-drying. Many cracks in freeze-dried samples could however be healed by re-wetting and air-drying. A crack which was about 10μm long and 300nm thick could, after re-wetting and then air-drying, could be closed by capillary shrinkage. The tail of the crack was not fully closed, but the crack-closure was so tight that the healed part could not be distinguished from the surrounding structure. Within some regions, the welding of gold ligaments was visible. The crystal orientation changed by about 48° across the boundary between two fused ligaments, thus confirming that a high-angle grain boundary had formed between two cold-welded ligaments, although the crack was transgranular before healing. It was concluded that capillary shrinkage produced appreciable plastic deformation within the structure during air-drying. De-alloyed material with a low relative density contained high-density native cracks. Dry samples exhibited a smooth fracture surface, while wet samples had an intergranular fracture surface.

Iron

Crack-healing in a single crystal of iron during heating was monitored[31] at the nanometre-scale by means of *in situ* transmission electron microscopy[32]. As the temperature increased from room temperature to 773, 973 and 1160K the morphology of the specimen changed considerably. After cooling to room temperature, the specimen partially recovered. There was a micro-crack ahead of the tip of the main crack. During heating, the micro-crack began to heal at 973K and was completely healed at 1160K. When the temperature was decreased to room temperature, the micro-crack did not reappear. In the case of a micro-crack which was near to the main crack, the micro-crack did not change during heating from 300 to 973K. When the temperature was increased from 973 to 1000K, the size of the micro-crack decreased gradually from 250 to 160nm. Upon increasing the temperature to 1023K, the size of the micro-crack further decreased to 70nm. After some 100s at 1023K, the micro-crack was completely healed. The average rate of crack-healing (table 4) was about 0.32nm/s. The self-diffusivity of α-Fe was known to be described by,

$$D(cm^2/s) = 2.0\exp[-28950/T] \text{ or } D(cm^2/s) = 118\exp[-35800/T]$$

The average diffusion coefficient between 973 and 1023K was $3.8 \times 10^{-13}cm^2/s$, and the total diffusion time was 400s. The self-diffusion distance was thus 123nm. This value was consistent with the crack-healing behaviour. It was concluded that the healing of cracks in α-Fe was due to self-diffusion.

Table 4. Crack-length and rate of crack-healing in α-iron

Temperature(K)	Crack-length(nm)	Time(s)	Healing-Rate(nm/s)
973	250	0	-
1000	160	162	0.28
1023	70	300	0.33
1023	0	400	0.35

An experimental and numerical investigation was made[33] of the high-temperature healing of micron-size intragranular micro-cracks in pure iron. Irregular penny-shaped micro-cracks were created by low-cycle fatigue, and the material was subjected to annealing in vacuum at 1173K. A penny-shaped micro-crack could evolve, via surface diffusion, into an isolated spherical void. Subsequent evolution caused doughnut-shaped pores to break up into a ring of spherical voids. The commercially pure iron (0.025wt%C) was fatigued, and then annealed (1173K, 5h, vacuum). Morphological changes in the internal cracks were then monitored. Intragranular micro-cracks alone were considered. A typical micro-crack had an irregular penny-like shape, and its morphological evolution could be described by the Nichols-Mullins model[34]. The micro-crack thickness, h, ranged from 0.2 to 3µm and the length/thickness ranged from 5 to 45, with L being the diameter of the crack. Because of the difference in chemical potential between the edge and the centre of the crack, a doughnut-shaped channel pore was expected to form along the outer edge of the crack via surface diffusion.

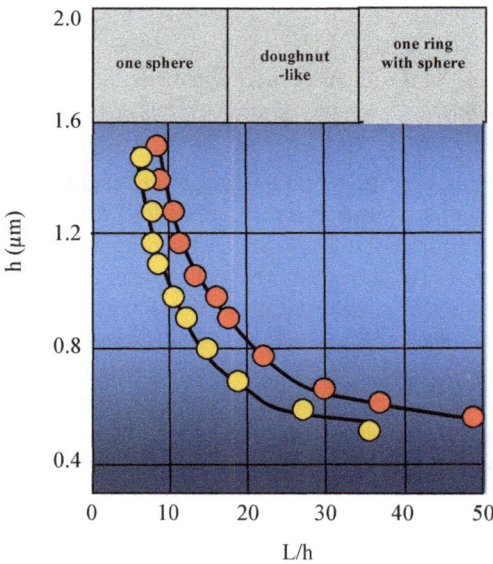

Figure 5. Size and aspect-ratio of penny-shaped micro-cracks in iron at 1173K, and their evolution into various profiles. Red: 2h, orange: 5h

The doughnut pore then broke up into arrays of spherical voids via Rayleigh-instability. A finite-element method was used to simulate the morphological evolution of a penny-shaped micro-crack under surface diffusion. The finite-element results showed that the penny-shaped micro-crack could evolve into an isolated spherical void, a doughnut-shaped channel pore and a spherical void surrounded by a doughnut-shaped cylindrical pore. The experimentally-observed void configurations were entirely consistent with the finite-element predictions. The time required for a penny-shaped micro-crack to complete the first stage of healing, as defined by the Nichols-Mullins hypothesis, could be calculated. Based upon the calculations, the initial dimension and aspect-ratio of a penny-shaped micro-crack to complete the first stage of evolution within 2 or 5h was such that the aspect-ratio was inversely proportional to h (figure 5). It was found that micro-cracks with an aspect-ratio of between 5 and 35, and an h of between 1.6 and 0.5mm, could complete the first stage of evolution in no less than 2h at 1173K. On the other hand, 5 hours were required when the aspect-ratio was between 10 and 45 and h was between 2.0 and 0.4mm. When the size and aspect-ratio of an internal crack had been experimentally determined, the first-stage evolution time could be estimated. Experimental processes for crack-healing could thus be designed on the basis of temperature, time and crack parameters: if the dimensions of the internal cracks could be estimated using the laws of crack growth, the morphological healing procedure could be used to eliminate cracks and restore properties.

Molecular dynamics methods were used[35] to investigate micro-crack healing in body-centred cubic iron. The temperature had a marked effect upon the micro-crack healing. Crack healing occurred within 6ps at 1173K. When the temperature was lowered, the rate of reduction in crack-size greatly decreased. The critical temperature for micro-crack healing in crystals with no pre-existing dislocations was about 673K. Defects such as dislocations and voids appeared during healing, and their positions continually changed. The simulation was relaxed for a long time at a temperature near to 0K. No defects were introduced into the initial configuration. A parameter was introduced in order to compare the healing process at various temperatures. It represented the minimum vertical distance between atoms on the top crack surface and those on the bottom crack surface. The parameter was equal to $1.526a$ in the initial configuration, where a was the lattice constant. Because the body-centred to face-centred cubic phase transformation occurred at 1185K and because the N-body constants for the former were not suitable for the latter, the model temperature was set to be 1173K. The time-step was $1 \cdot x \ 10^{-16}$s. The size of the crack decreased rapidly at this temperature. After 0.1ps, the comparison parameter was equal to $0.989a$. After 1.1ps, the crack had broken into two sections. After 1.8ps, only two micro-voids remained, at the positions of the 2 crack-tips. After 4.5ps, the crack was completely healed. The state of healing then remained stable. When the temperature was reduced, the rate of reduction of the crack-size markedly decreased. At 1073 K, the left-hand crack-tip was still not healed after 6.0ps. Neither crack-tip was healed at 973K. At 873K, the crack closed slightly but exhibited a lower rate of healing. During healing, atoms deviated from their normal positions, resulting in the presence of many defects such as dislocations and voids. The positions of the defects changed continuously and the distribution of defects was inhomogeneous. The defects aggregated near to regions where the original crack was located, and especially at the crack tips. In order to determine the critical temperature for crack-healing, the simulation covered temperatures ranging up from 273K in steps of 50K. Healing occurred when the comparison parameter was smaller than the cut-off radius of $1.300a$. At 623K, the parameter was equal to $1.304a$ after 6.0ps, and healing did not occur. When the temperature was 673K, the parameter was equal to $1.190a$ after 6.0ps and healing occurred. It was deduced that the critical temperature for crack-healing was about 673K. The critical temperature for micro-crack healing in body-centred cubic iron was lower than that in aluminium or copper, but there was no notable difference in micro-crack healing between body-centred and face-centred cubic crystals. During a real experiment, a pre-crack with a length of about 3.5μm and a width of 0.6μm, healing began at 973K. A long and narrow micro-crack which was about 3μm long and 0.2μm wide began to heal at 863K; 190K higher than the critical temperature in the simulation.

A molecular dynamics simulation was similarly used[36] to investigate the evolution of a nanoscale crack in body-centred cubic iron under conditions of heating or compression. An N-body potential and an embedded-atom method were used. The original nanoscale crack was modelled by removing atoms from the centre of the cell. The minimum vertical distance between the atoms of the top and bottom crack surfaces was defined to be d. The crack-healing process increased rapidly with increasing temperature. When the temperature decreased, the d-value of a healing crack did not change greatly but fluctuated within a narrow range, indicating that crack-healing was the result of iron atoms diffusing into the crack region and not of thermal stresses. A pre-compression pressure, applied under conditions of biaxial and uniaxial loading, could markedly promote healing and resulted in a more uniform defect distribution following healing. The evolution of a nano-crack in body-centred cubic iron under a compressive pressure was investigated. The crack was about 5.957×10^{-9}m long and more than $1.500a$ wide, where a was the lattice constant and was equal to 2.8665×10^{-10}m. The angle between the ($\bar{1}\bar{1}0$) slip-plane and the crack-plane was 60°. The width was greater than the cut-off radius of $1.300a$, meaning that there was no interacting force between the atoms on each side of the crack surface. The total number of atoms used in the simulation was some

13000. The x-axis lay along the central crack, the y-axis was normal to the crack-plane, and the z-axis lay along the [1$\bar{1}$2] direction. The size of the calculation-cell along the x-axis direction was $44a\sqrt{3}$, that along the y-axis was $24a\sqrt{2}$ and that along the z-axis was $a\sqrt{6}$. Each 3 layers of atoms along the x-axis was a cycle which was $a\sqrt{3}/2$ long. Each 2 layers of atoms along the y-axis was a cycle which was $a\sqrt{2}$ long and each 6 layers of atoms along the z-axis was a cycle which was $a\sqrt{6}$ long. The simulation of the crack behaviour under a pre-compressive pressure involved 2 steps, with a displacement boundary-condition and a fixed boundary-condition being imposed during the first and second calculation steps, respectively. During the first step, a compressive load was applied to the simulation cell and a displacement boundary condition was imposed along the x and y axes, with biaxial and uniaxial compressive loads being accounted for, respectively. The boundary-atom displacements were based upon the plane-strain linear-elastic displacement field in the solid-containing crack. The configuration at the end of the first step was then used as the starting configuration for the second step, age and a pre-compressive pressure existed in the simulation cell. During this second step, a high temperature was applied; starting at 523K and increasing in increments of 50K. Fixed boundary conditions were imposed along the x and y axes, thus ensuring that the locations of the boundary atoms remained constant during the calculation. Periodic boundary conditions were always imposed along the z-axis. The assumed initial velocity was taken from the Maxwell-Boltzmann distribution which corresponded to the given temperature, and was maintained constant by scaling the atom-velocities during the simulation. Equilibrium of the original configuration was attained after relaxation for a long period at a temperature close to 0K. There existed no defects in this original configuration. The magnitude of d was used to monitor the progress of crack evolution, with d being equal to $1.526a$ in the original configuration. Crack-healing became possible only when d was less than the cut-off radius of $1.180a$. The parameters of the N-body potentials for body-centred cubic iron were valid only below 1185K. Simulations were carried out between 273 and 1173K, in increments of 50K. The simulation spanned 6.0×10^{-11}s with a time-step of 1×10^{-16}s. There was a marked decrease in d during the first 1.0 a 10^{-12} and 2.0×10^{-12}s. It then increased, and fluctuated within a narrow range. The marked decrease could have resulted from instability of the calculations, and the simulation was again divided into 2 steps. During the first step, the environmental temperature had to be sufficiently low to eliminate the temperature effect. That is, 40K in the present case. The loading-rate of 2.5×10^{16}Pa$\sqrt{m}$/s was necessarily many orders of magnitude lower than that of a real loading. This showed that the existence of compression could both help to reduce the temperature required for crack-healing and also greatly promote crack-healing. The value of d fluctuated around the cut-off radius of

$1.180a$ between 523 and 673K. It fluctuated around $0.5a$ to $0.7a$ between 723 and 873K. The value of d fluctuated around $0.6a$ or less when the temperature reached just 573K. At 573 to 773K, the fluctuation-range did not generally fall with increasing temperature. But at 773K, for the configuration following uniaxial compression, d decreased rapidly to $0.6a$ after 1×10^{-11}s and then fluctuated around $0.3a$. No matter whether the simulation began from the original configuration, or from the configuration under pre-compression, cracks could be completely healed when the temperature reached 923K. During compression in the first step of the calculation, atoms deviated from their normal positions in some area. It was deduced that the existence of the defects triggered the crack-healing process. There were some defects in the crack-healing region and the surrounding area after 6.0×10^{-11}s at 1173K when starting from configurations until pre-compression. It was concluded that the existence of defects in the simulation cell following compression were helpful in producing a more uniform distribution of defects after crack-healing.

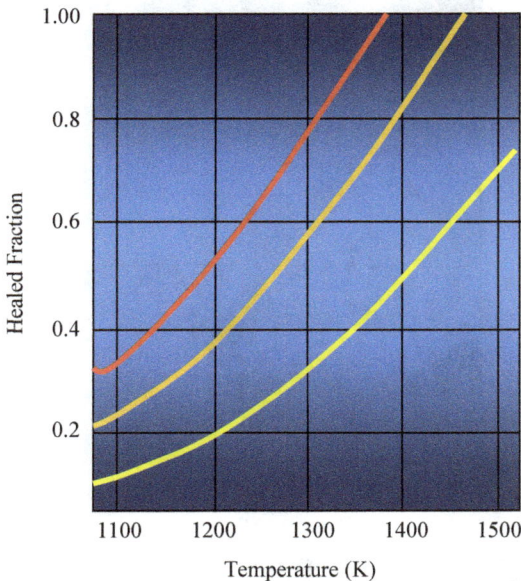

Figure 6. Healed-crack fraction versus temperature diagram for 34MnV steel. Red: 7200s, orange: 1260s, yellow: 60s

A kinetic model was developed in order to predict crack-healing as a function of temperature, pressure and time[37]. This model described the crack-surface topography in terms of a series of semi-spherical pores, and invoked a new diffusion healing-mechanism which involved grain growth. Plastic deformation, power-law creep and diffusion-controlled creep were also considered. A crack-healing diagram for Fe-0.35C-1.3wt%Mn (34MnV steel) was constructed, and the predictions of the model compared well with experimental data, with the critical healing time and pressure being determined by using the diagram.

Diffusion-controlled creep contributed little at high temperatures, because of grain growth. The geometry of the crack surface was modelled as an array of semi-spherical pores, with the radius of the sphere being equal to the crack-surface roughness. It was assumed that, when the opposing surfaces of the crack made contact, the two sets of spheres touched trough-to-trough to form micro-pores between them.

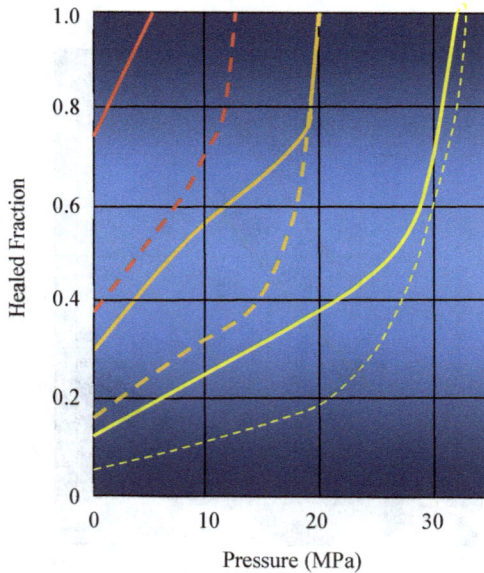

Figure 7. Healed crack fraction and pressure diagram for 34MnV steel. Red: 1473K, orange: 1273K, yellow: 1073K, solid lines: 1680s, dashed lines: 60s

This accurately reflected the shape of real fracture-dimples, where some of the micro-voids are near-spherical. It was further assumed that the entire crack surface was under hydrostatic pressure. Plastic yielding, power-law creep and diffusion-controlled creep was proposed to occur during healing. Deformation would take place when the effective pressure satisfied the yielding condition. Micro-creep also led to crack-healing at high temperatures, and this could be power-law, lattice-diffusion controlled and grain-boundary diffusion controlled. The lattice-diffusion controlled creep could be described by a Nabarro-Herring equation. The grain-boundary diffusion controlled creep could be described by a Coble creep equation. Lattice- and grain-boundary were combined into a unified diffusion mechanism. A crack-healing diagram could be constructed. The healing by yielding was time-independent, so the healed fraction which was due to plastic yielding represented the initial fraction of the time-dependent healing mechanism. The healing-rates of the power-law and diffusion-creep mechanisms were summed in order to give the total rate. The final healed fraction was calculated by integrating the total rate equation. In the current study, a new cracking healing kinetic model has been established. Crack-healing experiments performed on 34MnV steel and gave results which compared well with the model's predictions. A crack-healing diagram was able to predict the healed fraction quantitatively for various temperatures, pressures and healing-time (figures 6 and 7). A comparison of grain-growth and fixed grain-size models showed that the diffusion-controlled creep mechanism contributed little at high temperatures, due to the grain growth. The critical healing time and pressure could be deduced from two types of crack-healing diagram.

The healing of hydrogen-attacked cracks in low-carbon steel and Fe-2.25Cr-1wt%Mo steel was studied[38] by heat-treating split specimens in vacuum. This showed that crack changes were much smaller under conditions of pure heating, especially at the crack tips. The healing effect was closely related to the crack length; especially for shorter ones. The main mechanism being thermal diffusion, iron and carbon atoms had to diffuse at high speed in order that the plasticity deformation energy exceeded the surface tensile force energy. Phase transformations, and the stress-strain relationship, could have a positive effect.

A molecular dynamics study was made[39] of how the localization and transfer of excess atomic volume by structural defects affected the self-healing of nanosize cracks in single crystals of body-centred cubic iron under various loading conditions at room temperature. Deformation initially involved local growth of the atomic volume at crack tips. The crack-growth behaviour depended upon whether the excess atomic volume was transferred, by structural defects, from crack tips to the free surface or other interfaces. When an edge-crack was oriented, with respect to the loading direction, so that dislocations were not emitted from its tip, or twins alone were emitted, the sample exhibited brittle-ductile fracture. The transfer of excess atomic volume, by dislocations, from the crack tip prevented the opening of edge-cracks and was an effective healing mechanism for nanocracks. The fact that loading could lead to crack-healing, and increase the fracture resistance of a material was of practical value. Cracks in body-centred cubic iron could also be partially closed by heat and compression. These were supposed to promote a reduction in surface energy, and the healing of micro-cracks. Molecular dynamics simulations were here made of iron with micro-cracks at increasing temperatures. These showed that micro-crack healing was due to the diffusion of iron atoms into the crack region, rather than to thermal stresses. Additional compression of heated samples increased the healing. A transition from crack growth to healing was associated with dislocation-generation, twinning, recrystallization and phase transitions, due to an appreciable redistribution of atomic density. The phenomena were related to the formation of local regions of atomic volume, and to the transfer of this volume. The simulations showed that the formation of excess atomic volume during loading was governed by crack growth and by the emission of structural defects from the crack-tip region. Edge-cracks first grew in width and, the closer they were to the free surface, the greater was the gap between the crack faces during loading. The atomic volume in the tip region did not change until the crack began to open. The simulation results showed that the faces of the $(01\bar{1})[011]$ crack hardly moved between 20 and 33ps and the atomic volume remained essentially unchanged. During further loading between 35 and 93ps, the crack-width increased, with a slow increase in the atomic volume in that region. The variation in atomic volume was due to the emission of dislocations from the crack-tip region. The dislocations glided on the $\{112\}$ planes and carried away some of the excess atomic volume, so its expansion-rate was greatly reduced. Dislocation emissions from the tip region were related to the appearance of flat sections on the plot of excess atomic volume versus time. The escape of dislocations to the loaded surface led to the formation of steps. The simulation showed that dislocation-emission first slowed, and then decreased, the growth of tensile stresses. Changes in the excess atomic volume at the crack-tip, and in the tensile stresses, were closely related. The opening of the $(01\bar{1})[011]$ crack was thus suppressed by the transfer of excess atomic volume, by dislocations, from the crack-tip to the loaded surfaces. Atomic-bond breaking was possible only at a certain threshold value of the excess atomic volume. The simulation results agreed well with previous results on iron single crystals under mode-I loading. Simulation results showed that shear loading of a sample with a $(11\bar{2})[\bar{1}10]$ crack (table 5) was elastic at up to 6.5% strain. The rectangular incipient crack became diamond-shaped during loading and, upon reaching the yield-point, an edge dislocation was emitted from its tip. As a result of dislocation-emission, the volume of the crack decreased and the internal stresses relaxed. A crack in body-centred cubic iron was completely healed by heating to 1073K. It was assumed that heat-treatment and mechanical-treatment provided the driving force for minimizing surface energy and encouraging crack-healing. There was a periodic emission of dislocations from the crack-tip during loading when maximum stresses were reached. The emission period was governed by the strain-rate and the lattice period in the tensile direction. All of the dislocations generated glided on the (111) plane and, upon reaching the free surface, they produced a step. They left behind a small number of vacancies in the slip-plane. For complete healing of a $(11\bar{2})[\bar{1}10]$ crack of a given size, the shear strain had to be about 30%. A (011)[011] crack also healed via dislocation-emission from its tips under shear loading. The change in the nanocrack-volume was closely related to the internal-stress variation. Stresses in a single crystal of given orientation were concentrated in the crack-tips and in restraint-points on the free surface. During loading, edge-dislocations were also emitted from stress-concentration zones. The dislocations glided on {111} planes and left no vacancies behind. The thickness of the healed crack decreased after each dislocation-emission, its faces closed and became curved at the tips. Due the existence of rigid non-deformable regions, the dislocations could not reach the surface and remained in the bulk of the material. For a given sample-orientation and loading, the internal stresses increased during the entire loading period.

Table 5. Simulated cracks in molecular dynamics models of iron

Crack Location	Crack Orientation	Sample Orientation
edge	$(01\bar{1})[011]$	X [100], Y [01$\bar{1}$], Z [011]
edge	(010)[001]	X [100], Y [010], Z [001]
edge	(110)[001]	X [1$\bar{1}$0], Y [110], Z [001]
edge	$(11\bar{2})[\bar{1}10]$	X [111], Y [11$\bar{2}$], Z [$\bar{1}$10]
central	$(11\bar{2})[\bar{1}10]$	X [111], Y [11$\bar{2}$], Z [$\bar{1}$10]
central	$(01\bar{1})[011]$	X [100], Y [01$\bar{1}$], Z [011]

The healing of internal cracks in as-cast 30Cr2Ni4MoV steel was studied[40] following deformation. Internal cracks were produced by drilling, and the samples were then compressed to various degrees. Static and dynamic mechanical properties of the crack-healing zone were determined by means of room-temperature tensile tests and impact tests, respectively. Dynamic recrystallization and grain growth were the main factors which were involved in internal crack healing. When the forging ratio was 1.5, cracks at the corner of the void began to heal due to dynamic recrystallization. At a forging ratio of 2.0, dynamic recrystallization was complete and a centre crack was completely healed. The tensile properties of healed-crack zones were restored to be better than 95% of that of the base material. When the forging ratio was increased to 2.2, the elongation increased slightly while the yield strength decreased slightly. This suggested that grain-growth played a significant role in plastic recovery, while dynamic recrystallization played a large role in strength-recovery. The dynamic mechanical properties of the crack-healing zone gradually increased with increasing deformation. The maximum value of impact toughness occurred at a forging ratio of 2.0, and the recovery of impact toughness was greater than 96%. When the deformation was increased, the grains grew following dynamic recrystallization, causing the impact energy to decrease. When the forging-ratio was less than 1.8, the elongation of samples with voids was less than 1%; indicating that the plasticity of the material had not been restored. At a forging-ratio of 2.0, the elongation following crack-healing was 11.02% and the elongation of material without voids was 11.27%; a recovery-rate of elongation was 97%, indicating that the internal crack had disappeared and that the elongation-recovery was complete. When the forging-ratio was 2.2, the elongation following crack-healing was 11.17%; corresponding to that of the matrix. The reduction in area as a function of forging-ratio was essentially the same as that of the elongation. At a forging-ratio of 2.0, the reduction in area after crack-healing was 37.69% and the reduction in area of material without voids was 39.29%. The recovery-rate of the reduction in area was 96%. When the forging-ratio was 2.2, the recovery-rate of the reduction in area was 98%. This was because the grains continue to grow after recrystallization. On fracture surfaces, small equiaxed dimples appeared in healed areas. When the forging-ratio was increased to 2.0, the size and depth of the equiaxed dimples gradually increased. At a forging-ratio of 2.2, the fracture-morphology had changed little.

The evolution of internal crack-healing in 3 steels, including high-quality Fe-0.2wt%C structural steel, 30Cr2Ni4MoV steel and SUS304 stainless steel at high temperatures was investigated[41]. The existence of 2 crack-healing mechanisms was identified in all of the experimental steels. The crack-healing was controlled by atomic diffusion at temperatures below 1173K, and depended mainly upon recrystallization and grain-growth mechanisms at 1273K. Atomic diffusion provided material for recrystallization and grain growth in the crack-healing zone. Recrystallization led to rapid crack-healing, and grain growth promoted microstructural homogenization and elimination of the crack-healing zone. Crack-healing studies of 30Cr2Ni4MoV were performed at 900, 1000, 1100 and 1200C for periods of 5, 10, 15, 30 and 60min. Studies were similarly made of SUS304 stainless steel at 900, 1000, 1100 and 1200C for 1h. The behaviour of the crack-healing in the SUS304 was deduced from grain orientation and grain-boundary observations. After healing for 0.5h, the microstructure of the matrix was still homogeneous. With increasing holding-time, the distribution of ferrite in the matrix near to the crack-zone became more uneven. The pre-crack still existed and was unsegmented, while the crack-width steadily narrowed with increasing holding-time. This indicated that the crack-healing in 0.2wt%C steel had occurred at 900C. Ferrite alone was found in the crack-healing zones. A more uneven distribution of ferrite after healing for 2h indicated that the iron atoms of the matrix continually migrated towards the crack-healing zone and transformed into ferrite during the subsequent cooling process. Fine ferrite grains were not observed near to the original crack-gap, indicating that crack-healing in the 0.20wt%C steel at 900C, or below, depended mainly upon atomic diffusion and migration. Inspection of the crack-healing zone in 30Cr2Ni4MoV, after healing at 900C (1h) and air-cooling, revealed large bulges on the original crack face. This facilitated transformation of the original long crack transformed into intermittent micro-voids. Two micro-void bands formed on the sides of the original crack, along its length. Their widths were almost equal. The distances between the micro-voids at the edges of the micro-void bands and the original crack-centre were essentially identical. A pre-crack in the 30Cr2Ni4MoV steel healed at 900C, and the formation of micro-void bands indicated that crack-healing in steels at lower temperatures was dominated by atomic diffusion. The formation of the micro-void band was due mainly to a large amount of surface-energy, strain and lattice-distortion energy in the crack-zone. When nearer to the crack-face, these energy levels were higher. The combination of these factors generated a higher diffusion-rate at points close to the crack-face. The diffusion-rate difference led to the formation of the micro-void bands on each side of the original crack because, when atoms in the matrix near to the sides of the crack migrated to the crack-face, their places could not be quickly re-filled with new atoms and micro-voids resulted. The further away from the crack-face, the larger were the micro-voids in the crack-healing zone. Austenite-recrystallization aided rapid crack-gap filling, and grain-growth led to complete crack-healing. Crack-healing in 30Cr2Ni4MoV steel at higher temperatures was studied by holding at 1000C for 1h. Fine recrystallized grains were generated in the crack-healing zone. The crack-width at 1000C was much smaller than that at 900C, indicating that the

Materials Research Forum LLC
https://doi.org/10.21741/9781644903773

formation of recrystallized grains promoted rapid crack-filling. When the healing temperature was 1100C, the original crack-gap was again filled with a large number of fine recrystallized grains. This behaviour was identical to that of Fe-0.20wt%C at 1100C. Grains in the crack-healing zone at 1100C were much finer than those in the matrix at 900C, indicating that recrystallization had occurred in the crack-healing zone. After healing at 1200C for 1h, the microstructure of the crack-healing zone was indistinguishable from that of the matrix, proving that the original crack had been completely healed. Some newly-formed grains grew through the pre-crack gap. The depth and width of the centre meeting-line suggested that this line was effectively a new grain boundary and that the pre-crack had transformed into the new grain boundary. Similar phenomena occurred during crack-healing in Fe-0.20wt%C. It was concluded that two crack-healing mechanisms operated in Fe-0.20wt%C and 30Cr2Ni4MoV steel. The mechanism was atomic diffusion at temperatures below 900C, while the mechanism involved mainly recrystallization and grain growth at temperatures above 1000C. A study of crack-healing (900C, 1h, water-cooling) in SUS304 clearly showed that parts of the crack faces on both sides bulged markedly and that the original long pre-crack was separated into several short cracks. The bulges on the crack faces led to differing widths at various crack positions. Bulges on the original crack-face aided crack-healing. Micro-voids also formed in the matrix, near to the crack. After healing (900C, 1h), the crack morphology of SUS304 was similar to that of 30Cr2Ni4MoV. Recrystallization took place in the matrix of SUS304 (900C, 1h), and this was attributed to a 45% height-reduction during crack pre-setting. No fine equiaxed grains were found in the crack-healing zone along the crack length, indicating that internal-crack healing in SUS304 steel was also controlled by atomic diffusion and migration. Obvious healing occurred, and the original crack-gap had almost completely disappeared, while a centre dividing-line remained and some micro-voids were present along the line. There were massive fine grains on both sides of the crack, along the crack length. The fine grains proved that recrystallization had occurred in the crack-healing zone, and the centre dividing-line suggested that the growth-rates of recrystallized grains on both sides of the crack were equal. When the healing temperature was increased to 1100C, the number of fine grains decreased and this was attributed mainly to fine-grain growth or to the annexation of larger grains in the matrix near to the crack.

Gas-nitrocarburizing was used to heal cracks in metallic materials, especially 42CrMo (Fe-0.41C-1.06Cr-0.16wt%Mo) steel. The nitrocarburizing was performed in two stages, with decreasing healing temperature[42]. The optimum healing effect was obtained by treatment at 760C for 2h plus treatment at 550C for 4h. The maximum extent of healing was 63.68%. The healing process involved 2 stages: healing of crack tips at high temperatures and of crack opening at low temperatures. Volumetric expansion and filling of the nitrides formed contributed to a rapid healing of large-sized cracks. The healing efficiency was improved by decreasing the healing temperature. High-pressure gas nitrocarburizing was deemed to be a possible means for improving the efficiency and extent of healing. Artificial cracks were introduced into steel bars by heating at 940C for 0.5h and then quenching in oil. A single crack formed along the axial direction of the bar. The width of crack opening and tip were about 10 and 2µ, respectively. The nitrocarburizing was carried out in a mixture of NH_3 and organic gas at a pressure of 0.11MPa. It was performed in two stages. The material was first nitrocarburized at 760C for 2 to 4h and then at 670C or 550C for 2 to 4h before cooling in air. The crack-tips were almost fully healed by nitrides, and the healed morphology near to the crack opening varied with healing temperature. In some samples, no obvious nitrided layer was observed in the healed area, apart from some fine ferrite grains. This was attributed to the diffusion of iron from the matrix to the crack surface. In other samples, massive columnar nitrides grew from both sides of the crack surface and towards the crack-centre. The formation of columnar nitrides was beneficial for the rapid healing of wide cracks. Upon decreasing the temperature to 550C, the healed area then had a denser appearance. The morphology of the healed area affected mechanical properties following crack-healing. It was concluded that the principal healing mechanism was the filling-with and connection of nitrides in the crack gap. The healing process could be divided into two stages. In the first stage, healing in crack-tips occurred at a high temperature. For this, a diffusion layer was available at the cracked surface due to the diffusion-dominated nitrocarburizing at austenitic temperatures. The diffusion layer consisted mainly of Fe-N austenite and ferrite. In the second stage, nitrogen atoms could continuously permeate into the narrow cracks under the driving force provided by the concentration gradient between the crack-tips and opening. Nitrogen which was dissolved in austenite and ferrite led moreover to a volume expansion which could directly reduce the crack width. The degree of reduction depended upon the nitrogen concentration in the austenite and ferrite. Austenite with 10at% of nitrogen in solution at 760C could provide a maximum volume expansion of 6.16%. The crack-tips therefore healed first at the high temperature. Healing in crack openings depended upon phase transitions and nitride formation at low temperatures. The filling efficiency was given by the relative volume expansion. The ε-Fe_2N, ε-Fe_3N, γ' and $Fe_{16}N_2$ nitrides provided a volumetric growth of 24, 17.8, 15.6 and 9.2%, respectively. Rapid growth of the columnar layer of nitrides occurred at 670C, further accelerating the healing-rate of the cracks; especially in the crack openings. Crack-healing ceased when the crack-opening was entirely full of ε-nitrides, leaving some unhealed holes beneath the crack-openings. It was noted that the most general means for increasing the degree of healing was usually to increase the healing temperature or lengthen the holding time. In the present case, the key factors in rapid crack-healing were instead phase transitions and the growth of nitrides, rather than atomic diffusion. The healing efficiency and degree of healing could here be achieved by decreasing the healing temperature. The greatest barrier to healing cracks by using gas-nitrocarburizing was to ensure the continuous permeation of nitrocarburizing agents into the narrow cracks. The diffusion of nitriding elements into narrow cracks depended upon the mean-free-path of molecules.

FeCoCrNiAl

An investigation of crack-healing mechanisms in $FeCoCrNiAl_{0.5}$ high-entropy alloy under ion irradiation was performed by means of molecular dynamics simulations[43]. An analysis of the generation and recombination of point defects during the formation of overlapping collision-cascades revealed the crack-healing mechanisms which occurred in the high-entropy alloy. Interstitial defects which were generated within the core of a cascade, during the initial collision event, diffused to the crack surface and led to crack-healing during subsequent recrystallization. The corresponding vacancies also accumulated and formed large vacancy clusters that then generated stacking faults and complex dislocation networks which were distributed around the healed crack. Upon increasing the number of overlapping cascades, the defect recombination-rate increased and the phase stability was further improved. Unlike conventional alloys, these materials generally comprise 5 or more principal elements in almost equal proportions, and tend to form simple face-centred cubic and body-centred cubic solid-solution phases. This ultimately increases defect recombination and annihilation, and contribute to a self-healing effect which imparts a marked radiation-tolerance. During bombardment, atomic defects in these alloys require high formation-energies, thus promoting rapid recombination and aiding defect annihilation. Ion-bombardment can modify and possibly heal irradiation damage and is thus very relevant to the microcrack-healing behaviour. Studies of pure metals have shown that collision cascades can heal nanocracks, thus providing a guide to understanding similar mechanisms in high-entropy alloys. The study of the present alloy concentrated on the irradiation behavior of pre-cracked samples. The formation and evolution of point defects in cracked and defect-free material was examined. Following the first collision cascade, there were marked changes in the microstructure of the cracked samples. The heat-wave which was generated by the primary knock-on atom caused localized melting in the cascade-core, provoking a phase transition from face-centred cubic to amorphous. Irradiation-enhanced diffusion drove atoms towards the crack, where they rapidly accumulated during the thermal spike and filled the crack. The temperature profile at the cascade core indicated a rapid temperature rise to far above the melting point (2060K) of the alloy. This was followed by a sudden decrease to room temperature. This thermal relaxation promoted local atomic rearrangements and partial recrystallization. Crystallization occurred within the cascade core, and reflected a crack-healing which was accompanied by stacking-fault formation. During structural relaxation, the fraction of atoms with face-centred cubic and hexagonal close-packed structures increased at the expense of amorphous phase and reached a stable state after some 40ps. The resultant filling of the crack increased phase-stability and self-healing. In spite of this healing, the presence of many residual vacancies near to the crack led to the formation of stacking faults. When compared with defect-free samples, cracked samples exhibited more stacking-faults following an initial cascade. In the case of overlapping cascades, the stacking-faults persisted and reduced defect mobility during the relaxation stage, thus leading to a lower recombination rate. A quantitative analysis of Frenkel-pair generation and recombination in the case of overlapping cascades also clarified the atomistic mechanisms of crack-healing. The enhanced self-healing which was observed was closely related to the preferential absorption of interstitial atoms by cracks.

Fe-Cr-Ni

The corrosion of samples of 800H alloy, with aluminium contents of 0.02, 0.13 or 0.34% was investigated[44] at 1000C in air or in a purely nitriding atmosphere. Investigations concentrated on the corrosion processes which occurred at a growing creep-crack in air. A critical role was played by an oxide which formed in the crack. At the beginning of creep-crack growth, a dense chromia layer protected the metal around the creep-crack from internal nitridation. An increasing destruction of the oxide by deformation around the crack tip, and subsequent oxide crack-healing, led to chromium depletion in the metal and to the formation of a porous iron-rich oxide in the creep-crack. It was proposed that, due to oxidation and impeded gas-exchange in the porous oxide, the atmosphere within the creep crack became depleted in oxygen and the nitrogen partial pressure approached unity. Following an incubation time, the corrosion conditions in the creep crack thereby came close to those of a pure nitriding atmosphere; thus explaining the severe internal nitridation.

The healing of cracks in polycrystalline Fe-18Cr-9wt%Ni (SA508) samples was studied[45], for various loading conditions, by means of molecular dynamics simulation. This indicated that samples which were subjected to compression possessed a greater ability to resist plastic deformation whereas a shear stress promoted plastic flow. Crack closure and healing occurred under compression via dislocation-dominated plastic deformation; the crack-length decreased and the crack-tips extended along grain boundaries because of the higher stress under shear loading. Dislocation emission, slip, and interaction with cracks and grain boundaries contributed to the plasticity of specimens under compressive loading. Grain-boundary slip, grain-rotation and twinning were possible plastic-deformation mechanisms under shear loading. Under a shear stress, dislocation-emission from the crack-tip led to crack-closure via a dislocation-shielding effect and atomic diffusion. Crack-healing was known to depend largely upon the crystallographic orientation and the direction of external loading. High temperatures aided crack-healing because atomic diffusion occurred more readily. Compression, whether biaxial or uniaxial, promoted crack-healing and led to a more uniform distribution of defects following healing. Healing and recovery of the impact properties of SA508 steel were studied for various deformation-modes and heat-treatments. It was noted that the recovery of impact properties by multi-pass thermal deformation was lower than that produced by uniaxial compression at 950 and 1050C. Crack-closure and healing occurred under compression, due to dislocation-dominated plastic deformation.

Solidification-cracking was studied[46] in steel/copper-alloy structures, with SS316L (Fe-17.5Cr-12.5Ni-2.2Mo-2.0wt%Mn) and H13 (Fe-5.2Cr-1.6Mo-1.1wt%V) being deposited onto copper-beryllium substrates by using laser directed-energy deposition. Solidification cracks were observed on the steel side, with a downward trend occurring as the dilution-ratio increased. Crack-free single-layer SS316L specimens were obtained when the dilution-ratio was greater than 44%. The SS316L and H13 had a similar crack morphology and tendency, thus indicating that the cracking susceptibility of these materials was governed by the iron-copper phase diagram. That is, the copper concentration in the steel affected the cracking susceptibility via the solidification temperature-range and the amount of peritectic liquid. The cracking tendency increased markedly when the copper concentration was greater than about 9wt%. This was further driven by the wide solidification-temperature range and the limited final amount of liquid. At higher concentrations, the cracking tendency continually decreased as the copper concentration increased until a non-cracking region was reached when the copper concentration was 50wt% or more. The existence of a wide solidification range led to a long narrow liquid channel being present in the final stages of solidification, thus promoting crack formation. A larger final amount of liquid increased the back-filling effect, thus aiding crack-healing and reducing the cracking tendency. The cellular and columnar dendrite-microstructures which were typical for low copper concentrations increased the cracking tendency because thin elongated intercellular and interdendritic films were weak and susceptible to crack initiation. At higher copper concentrations, spinodal decomposition could instead produce a finely-distributed, grid-like arrangement of copper-rich phases within the steel matrix. This inhibited crack nucleation.

Three samples of additively-manufactured 316L/TiC stainless-steel composites were prepared[47] by using laser-based directed-energy deposition. One sample was crack-free, while the others contained cracks. Electropulsing was used to heal the cracks by using various current-densities, and 75 pulses. Cracks with widths of 0 to 200µm could be healed in this way, and the microstructure in the healed region consisted of refined fractured carbide particles and re-solidified fine austenite grains. Healed regions thus had fine TiC particles within the re-solidified zone, with a width of 50 to 200µm. Some fractured TiC particles were also found in, or adjacent to, the healed regions. Healed zones also contained fine re-solidified austenite grains as well as fine TiC particles. Due to current deviations and concentrations around defects and cracks, a temperature-field was generated and led to melting and to compressive stresses. After a single electropulse, the electric current concentrated at the crack tip. The average temperature rise could be calculated using the equation, $\Delta T = j^2 t/(cd\sigma)$, where σ was the electrical conductivity, d was the density, c was the specific heat, t was the pulse-duration and j was the current-density. So, assuming that j = 2.5GA/m^2, d = 6.5g/cm^3, t = 0.1ms, σ = 70.1 x 10^{5}A/Vm and c = 0.57J/gk, the maximum temperature-rise following a single pulse would be 241C. The 316L matrix near to the crack could melt where the temperature reached the melting point. High compressive stresses split large brittle TiC particles into small particles. The molten 316L matrix and fine TiC particles then mingled. Due to the rapid solidification, fine austenite grains formed. A new crack-tip was generated by one pulse, and further healing was repeated as before. Due to the reduced crack area following the first pulse, a higher current intensity was required for further electropulse healing. The high cooling-rate and steep thermal gradients generated high levels of strain. Following treatment using multi-pulses, the strains vanished. The present method was expected to be able to heal defects or cracks in many additively manufactured metal-matrix composites.

Magnesium

Mg-Gd-Y

A new method was proposed[48] for healing micro-cracks in Mg-6.84Gd-5.06Y-0.62Zn-0.96Zr-0.04Mn-0.02wt%Fe tubes by using high-density eddy-current pulses. By inducing electromagnetic currents within a copper coil connected to a pulsed power-source, a current-density of greater than 5 x 10^{9}A/m^2 and of short duration was generated in tubular specimens of magnesium alloy. Micro-cracks were clearly healed and the mechanical properties were improved. During internal treatment, the crack-healing was attributed to thermal stresses around the micro-crack tips, to softening or melting of the metal near to the micro-crack tips and to the squeezing effect of the Lorentz force. Compressive radial stresses and tangential stresses which were induced by the Lorentz force contributed to more efficient crack-healing and to improved mechanical properties. During outer treatments, the process could heal micro-cracks without direct contact with the tubular specimens and this was not limited by the length of the specimen. Following spinning deformation, micro-cracks were detected on the cross-sections of tube specimens and this was attributed to the reduced ductility of such rare-earth containing magnesium alloys and their increased strength. The micro-crack morphology of specimen A2 (table 6) was largely unchanged following 5 pulse-cycles. After 10 cycles, localized crack-closure was seen in specimen A3, with micro-cracks narrowed or fully closed. This trend in healing was more obvious in specimen A4, after 15 cycles. Detailed monitoring of specimen A4 every 5 cycles confirmed the progressive narrowing and closure of micro-cracks, thus proving the treatment's effectiveness in healing damage in magnesium-alloy tubes. A similar trend was found in the case of specimen A5, which was treated by using an inward-discharge scheme at 6kV. There was gradual micro-crack closure with increasing number of cycles. Tube specimens which were treated in the inward-discharge configuration exhibited a comparable micro-crack healing behaviour. Following 5 cycles at 9kV, partial crack-closure was observed in specimen B2. By the tenth cycle, specimen B3 exhibited a marked crack-reduction, and nearly all of the cracks in specimen B4 had healed after 15 cycles. Specimen B5, after 15 cycles of outward-discharge treatment at 9kV, also exhibited healing but not as much as B4. In order to judge the mechanical behaviour, uniaxial tensile testing was performed.

The average yield-strength of group-A specimens increased with increasing number of cycles, Specimen A1 (as-spun) had a strength of 170.83MPa, while A2, A3 and A4 had strengths of 182.37, 194.01 and 199.13MPa, respectively. These were increases of 6.75, 13.86 and 16.56%. The ultimate tensile strengths of A2, A3 and A4 exceeded that of A1 (280.91MPa) by 5.23, 10.66 and 12.76%. The elongation enjoyed particularly marked gains, from 10.37% for A1 to 12.68, 15.03 and 15.87% for A2, A3 and A4. Specimen A5 (inward-discharge, 15 cycles, 6kV) offered better mechanical properties than did A4 (outward-discharge, 15 cycles, 6kV) and the yield strength increased to 202.12MPa, the UTS to 298.47MPa and the elongation to 16.11%; all slightly better than those of A4. In group B, the yield-strength increased from 173.73MPa for B1 to 199.55, 212.41 and 215.23MPa for B2, B3 and B4. The UTS increased from 263.02MPa for B1, by up to 20.02% after 15 cycles. The elongation could increase by up to 62.77%. Specimen B5 (outward-discharge, 15 cycles, 9kV) also exhibited improvements, with a 215MPa yield strength, a 311.33MPa UTS and a 15.35% elongation. The results thus confirmed that micro-cracks in as-spun magnesium-alloy tubes could be healed by eddy-current pulse treatment, with improved healing and properties being related to the number of cycles and to the discharge voltage. Excessive cycling or high voltages could however lead to local melting. The precise healing mechanism involves thermal stresses around micro-crack tips which were caused by the eddy currents, localized melting or softening plus Lorentz-force induced compression. Inward-discharge schemes produced higher eddy-current densities and led to greater thermal stress and more effective crack-closure. The Lorentz-force compressive radial and tangential stresses further improved healing. In the case of outward discharge, tensile tangential stresses tended to counteract the healing process.

Table 6. Treatment conditions of eddy-current pulse-treated Mg-Gd-Y alloy

Specimen	Scheme	Voltage(KV)	Treatment Cycles
A1	outward	6	0
A2	outward	6	5
A3	outward	6	10
A4	outward	6	15
A5	inward	6	15
B1	inward	9	0
B2	inward	9	5
B3	inward	9	10
B4	inward	9	15
B5	outward	9	15

Continuous eddy-current pulse treatment was combined with heat treatment in order to heal micro-cracks in spin-formed Mg-6.84Gd-5.06Y-0.62Zn-0.96Zr-0.04Mn-0.02wt%Fe tubes[49]. All of the micro-cracks in the various tube specimens were essentially healed after continuous eddy-current pulsing for up to 15 cycles. Treatments which involved fewer cooling intervals provided a better healing effect and increased the strength and elongation of the alloy tubes. Following aging, the strength improvement of eddy-current treated specimens was more marked than that of specimens without that treatment. The decrease in elongation of eddy-current treated specimens was less evident than that of non eddy-current specimens. This was due to the segregation of rare-earth elements to the crack surfaces. Following solution-treatment, the strength-reduction and ductility-improvement of eddy-current treated specimens were more evident than for non eddy-current treated specimens. This was due to a marked decrease in the dislocation-density of eddy-current treated specimens. Narrowed cracks which were produced by eddy-current treatment, and the segregation of precipitates in the vicinity of the micro-crack surface during aging, explained the maximum strength which was exhibited by eddy-current treated as-spun specimens after aging. Some specimens were not subjected to pulsing, while others underwent continuous eddy-current treatment for up to 15 times at a discharge voltage of 7kV. The voltage and treatment-time were selected so as to avoid local melting. When subjected to pulsing 5 times, micro-cracks on the cross-section of the tube tended to close in some places while, cracks in adjacent areas also exhibited a strong tendency to healing. With increasing pulsing time, more and more micro-cracks closed or markedly narrowed. Most micro-cracks had healed after 15 times. The healing was attributed to thermal compressive stresses and to softening/melting near to micro-crack tips, caused by detoured eddy currents and by the squeezing effect of Lorentz forces. Apart from the micro-crack morphology, the microstructures of spin-formed specimens did not fundamentally change because of the relatively low temperatures which were induced by the treatment. The average yield strength, average ultimate tensile strength, average elongation and average hardness of as-spun specimen-A1 (table 7) was 209.5MPa, 313.3MPa, 10.2% and 91.3HV, respectively. The corresponding properties of specimen A2 were 220.5MPa, 330.74MPa, 12.4% and 94.5HV. The properties of specimen A3 were 224.4MPa, 336.2MPa, 13.3% and 97.2HV, respectively. Following 15 treatments, the properties of specimen A4 were 236.3MPa, 343.2MPa, 14.1% and 100.7HV, respectively. The improvements were attributed to micro-crack healing. In order to improve the mechanical properties further, heat treatment was applied to the tube samples following pulsing. This led to the yield-stress, UTS, elongation AE and AH of specimen C1 being 209.1MPa, 314.2MPa, 10.2% and 91.3HV, respectively. The corresponding results for B1 were 227.2MPa, 329.1MPa, 12.5% and 94.5HV. Following solution-treatment, the results for B2 were 151.5MPa, 247.1MPa, 20% and 86.6HV respectively. Following solution/aging treatment, the results for B3 were 211.2MPa, 335.0MPa, 16.4% and 92.3HV. Those for B4 were 261.3MPa, 365.2MPa, 11.1% and

126.2HV. Spin-formed tube samples in group C, without eddy-current treatment were also heat-treated. This affected the mechanical properties in a similar manner to those of pulsed specimens, but the properties of C-type specimens were slightly lower than those of B-type specimens. Following multiple pulsing treatments the micro-cracks were very narrow, and the filling of precipitates was expected to promote the healing of micro-cracks more effectively group-B specimens. The improvement in mechanical properties which was caused by aging was more obvious for B3 and B4 (pulse-treated) than that of C3 and C4 (without pulsing). The microstructure of as-spun specimens without pulsing had a low dislocation density, due to the occurrence of sufficient dynamic recrystallization during hot-spinning. The grain-size markedly increased after solution-treatment and resulted in the obvious decrease in strength. During subsequent aging, the precipitated phase appeared in the matrix and segregated near to the crack surface, leading to the strengthening of specimens. In pulsed specimens, the micro-cracks were narrowed and the dislocation density was increased by several pulsing treatments. Following solution treatment, the grain size of pulsed specimens also increased markedly but the dislocation-density decreased. The strength fell more obviously than did that of specimens without pulsing. Subsequent aging led to the segregation of precipitated phases around the micro-cracks, thus narrowing them and rendering them more easily filled by precipitated phase and thus healed. A greater strength improvement occurred in pulsed specimens after solution/aging treatment. When aging alone was applied to specimens, precipitate-segregation occurred around the micro-cracks but the relatively low aging temperatures made it difficult to affect the grain-size and dislocation-density. Due to the narrower micro-cracks existing after pulsing, the strength increase of pulsed specimen was more obvious than that of as-spun specimens, due to the greater micro-crack healing-effect. Specimens with or without pulsing offered a greater strength-improvement than did solution-treatment and aging, respectively.

Table 7. Treatment schedule of magnesium alloys

Specimen	Pulsing Time	Heat-Treatment	Grain-Size(μm)
A1	-	-	-
A2	5-5-5	-	-
A3	7-8	-	-
A4	15	-	-
B1	5-5-5	-	12.5
B2	5-5-5	solution	45.3
B3	5-5-5	solution/aging	49.5
B4	5-5-5	aging	13.6
C1	-	-	-
C2	-	solution	47.8
C3	-	solution/aging	50.8
C4	-	aging	12.9

Mg-Y-Ni

The hot-tearing tendency and crack-arrest mechanism of long-period-stacking ordered reinforced Mg-4Y-xNi alloys, where x was 0.5, 1, 2, 3 or 4wt%, were investigated[50] by using T-shaped constrained rod casting moulds. All of the alloys, apart from Mg-4Y-4Ni, suffered from a high hot-tearing susceptibility. With increasing nickel content, the crack-volumes at the hot-spots of the castings were 0.033cm^3, ∞ (fractured), 0.062cm^3, 0.020cm^3 and 0.004cm^3, respectively. The effect of the nickel content upon the susceptibility was characterised by a hot-tearing initiation temperature: 514, 593, and

436C, respectively. The Mg-4Y-4Ni alloy exhibited no detectable initiation temperature and had the lowest susceptibility. The Mg-4Y-1Ni alloy, with a columnar grain morphology, suffered hot-tearing at a stress evolution rate of 4.3N/s, due to localized stress-concentration during the early stages of solidification, corresponding to a critical load of 10.2N. The presence of fine grains accelerated stress-evolution in Mg-4Y-2Ni up to 14.6N/s. The increased liquid-film thickness, and intergranular bridging, delayed the initiation of hot-tearing. The increase in the residual liquid content increased the crack-healing ability. The volume fractions of the secondary phase in the alloys were 7.7, 14.9, 17.4, 20.2 and 24.9% with increasing x. A higher content could improve the crack-healing ability of residual liquid following the onset of hot-tearing initiation. The width of the long-period-stacking ordered phase in the 3 and 4wtNi alloys was greater than that in the other alloys. The greater ubiquity and size of that phase greatly increased the grain-boundary strength and thereby resisted solidification-shrinkage stresses. As the nickel-content increased, complete fracture occurred at the hot-spot of the 1wt% alloy and this implied an increase in the hot-tearing susceptibility of the alloy. A limited degree of crack-healing was still observed near to the fracture site but, due to early crack-initiation and rapid propagation, it was difficult to inhibit hot-tearing with a low content of residual liquid. An increase in residual liquid increased the ability of the alloy to feed shrinkage, and thus heal cracks. The width and depth of hot cracks decreased and no penetration-cracks formed. The hot-tearing volume of the alloys first increased and then decreased as the nickel-content increased. The lower ordered-phase content of the nickel-free alloy was insufficient for crack-healing and hindering propagation. In the 2wt% alloy, more traces of healing existed at the hot spot but it was difficult to resist cracking at a high propagation rate. With increasing healing and bridging effects, primary-crack propagation in the 3wt%Ni alloy was hindered and many fine traces of healing were observed, from the end of the crack to the centre. Only small partial cracks therefore formed. The 1wt%Ni alloy suffered from the highest hot-tearing susceptibility, due to complete fracture at the hot-spot of the casting. The 4wt%Ni alloy had the smallest hot-tearing volume and the lowest hot-tearing tendency, as the cracks were arrested at the surface. The hot-tearing initiation point was a critical point at which the solidification-shrinkage stress was in equilibrium with the tensile strength. When this point was passed, solidification entered a zone of high hot-tearing tendency. The contribution of a liquid film to the grain-boundary strength was much lower than that of intergranular bridging, both of which were very liable to stretching and tearing. Non-solidified residual liquid had a healing effect upon cracks but, as the liquid was absorbed by cracks, its reduced presence weakened bridging. An increase in nickel content increased the content of residual liquid as well as the thickness of the liquid film. In the 3 and 4wt%Ni alloys, an increase in the amount of feeding and ordered-phase bridging delayed or inhibited hot-tearing initiation and led to a marked increase in the final load. In the 1wt%Ni alloy, with a high hot-tearing initiation temperature, crack-healing led to enrichment of the ordered phase in localized regions and latent-heat release. This weakened the overall softening effect of the ordered phase upon the alloy structure and reduced its capacity to relieve stress-accumulation. The higher precipitation temperature of the ordered phase in the 2wt%Ni alloy led to a stronger bridging effect and reduced the hot-tearing initiation temperature. Although residual liquid can heal cracks quickly after hot-tearing initiation, secondary hot-tearing began due to the faster evolution of stresses which were greater than the grain-boundary strength. In the 3 and 4wt%Ni alloys, the higher latent heat of solidification release reduced shrinkage of the alloy structure and, together with an improved feeding capacity and ordered-phase bridging, reduced the hot-tearing susceptibility of those alloys. The freezing-range of the alloy widened with increasing nickel content, a factor which is usually considered to impair the hot-tearing resistance of an alloy. In multi-component alloys, it is however difficult to judge the susceptibility properly on this basis. It was suggested that hot-tearing initiation was due mainly to an insufficient strength of the alloy structure failing to resist shrinkage stress, together with a limited healing capacity of the residual liquid. When the void-volume increased due to crack propagation and the residual liquid was unable to provide timely healing, due to obstruction of the healing channel, this led to an abrupt decrease in local pressure and to the formation of a zone of negative pressure. In the mushy zone however, due to the complexity of the dendritic network and the narrow channel, liquid healing was easily delayed or frustrated due to a large flow-resistance. A local so-called vacuum-effect then rendered the pressure at the crack-tip appreciably lower than that in the surrounding area and set up a pressure gradient. Due to this gradient, the negative-pressure zone draws in any surrounding residual liquid which has not fully solidified. When the reserve of residual liquid is sufficient it can heal the crack. Small secondary cracks around the primary crack are also easily healed. The width and number of traces of residual-liquid healing increased with increasing nickel content. This implied that the increase in residual-liquid content improved the ability of the alloy to heal larger secondary cracks. This healing effect was most marked in the case of the 3wt%Ni alloy. Microstructures showed that the traces of cracks healed by the ordered phase and Mg_2Ni eutectic were the widest and most numerous and were continuously distributed. Healed cracks persisted in the alloy as defects, rather like macro-segregation, and offered a higher crack-resistance than before healing, thus preventing further crack-propagation. Such healing exerted a weak inhibitory effect upon hot-tearing initiation however, because the cracks had already caused load-release before healing. The 3wt%Ni alloy therefore still suffered

Materials Research Forum LLC

https://doi.org/10.21741/9781644903773

from a relatively high hot-tearing susceptibility. In the case of the 4wt%Ni alloy, there were only a few healing-related traces of residual liquid in the microstructure near to the hot-spot and the width of the healing traces became narrower. In this alloy, the solidification-heat of residual liquid relieved stress-development and weakened the stretching-effect of stresses upon the dendritic structure. The solidification of residual liquid at grain boundaries meanwhile led to the formation of intergranular bridges which strengthened the grain boundaries. Ordered phases which enjoyed a certain orientational relationship with respect to the α-magnesium matrix, were able to withstand higher stresses before fracture and could thus greatly increase the strength of the dendritic structure. The higher number of intergranular bridges composed of the ordered phase accelerated grain-boundary strengthening, and slow-to-develop stresses could barely cause greater tearing of the dendritic structure. The 4wt%Ni alloy thus exhibited fewer and finer healing-traces and suffered from the lowest hot-tearing tendency. Unobstructed filling-channels improved the intergranular liquid-feeding capability and prevented the formation of films with voids, which were very likely to act as nucleation sites for hot-tearing. It was known that fluidity tended to be inversely proportional to the freezing range and it was found that, apart from 4wt%Ni, the flow-lengths of the alloys followed that trend (table 8). During solidification, constitutional undercooling provoked the nucleation of crystals in the liquid phase which then impeded liquid flow. Alloys with longer freezing ranges were known to form crystals with irregular growth surfaces, again greatly reducing the fluidity of the alloy. Fluidity could also be related to precipitation of a secondary phase. In the case of the 1 and 2wt%Ni alloys, precipitation of the ordered phase occurred earlier than in the nickel-free alloy, thereby increasing the solid fraction of the alloy at a given temperature and impeding the flow of residual liquid. Subsequent solidification-shrinkage of confined residual liquid then promoted the formation of localized voids. These could not be removed by liquid-feeding and instead had to be eliminated by solid-state processes. When precipitation of the ordered phase was insufficient, any grain-boundary strengthening was limited and rendered the boundaries incapable of withstanding the stresses produced by the solid-state processes. This led to grain-boundary fracture and hot-tears. This accounted for the common cracking and high hot-tearing tendency of the 3wt%Ni alloy, but its superior healing-effect impeded further crack propagation. In 4wt%Ni alloy, a further lessening of fluidity did not cause it to suffer the lowest hot-tearing tendency. An increased precipitation of the ordered phase, which increased the ability of grain boundaries to resist stress, was suggested to be one of the main reasons for a reduced hot-tearing susceptibility. Following crack-initiation, any remaining non-solidified liquid flowed towards the crack, covered its surface to form wrinkles and thus partially healing it. In the 4wt%Ni alloy, just a few filamentous traces and smooth fracture surfaces were observed. The higher content of residual liquid improved intergranular feeding, eliminated void-formation and prevented severe hot-tearing during a dendrite-separation stage.

Table 8. Filling-lengths of casts of
Mg-4Y-xNi alloys as a function of x

x(wt%)	Filling-Length(mm)
0.5	610
1.0	414
2.0	318
3.0	309
4.0	282

Molybdenum

Hot isostatic pressing was used[51] to reduce pores and cracks in material which was produced by selective laser melting. When the applied linear energy-density was greater than 0.48J/mm, the samples could contain continuous molten tracks. When the applied volumetric energy density ranged from 250 to 280J/mm^3, small cubic selective laser-melted samples could have relative densities that were higher than 99.7%. In the case of large bulk laser-melted samples with a volumetric energy-density of 259J/mm^3, the addition of 0.9wt%La_2O_3 could reduce the porosity from 0.76 to 0.63%. Following hot isostatic pressing, the highest relative density of 99.6% was observed for large bulk samples with La_2O_3 additions. Cracks were observed only at the surface, none being found in the interior. After adding La_2O_3, the cracks at the surfaces were narrowed and the crack number-density was reduced more than two-fold. The treatment was concluded to be an effective means for both reducing the number of pores and for healing cracks in selective laser-melted material. During hot isostatic pressing, the material was subjected to high temperatures and pressures for some time. The temperature was chosen so as to reduce the yield strength of the material to below the applied pressure and thus permit plastic flow of close pores and cracks.

Nickel

The healing of cracks in metallic materials is difficult due to the sluggish atomic mobility in the solid state at room temperature. A new crack-healing approach[52] was based upon an electrochemical process in which metallic ions in an electrolyte were used as a healing agent. Pure nickel sheets with through-thickness cracks were used as a test-bed. Cracks with sizes in the micrometre range, or larger, were healed by the electro-treatment. The electro-healing began with the vertical epitaxial growth of healing crystals from the original crack surfaces, followed by the lateral growth of healing crystals which mutually bonded at the atomistic level. Tensile testing showed that healed samples possessed a tensile strength which was comparable to that of virgin samples, and tensile ductility was found for samples which were 100µm in thickness. The healing efficiency ranged from 96% to 33% with increasing sample-thickness. This was related to the fraction of fully-healed region, and to the strength difference between the substrate and the healed crystals. In detail, polycrystalline nickel plates with a purity of 99.84wt% and grain-sizes ranging from 100 to 150µm were used. The healing solution generally consisted of $NiSO_4 \cdot 6H_2O$, $NiCl_2 \cdot 6H_2O$ and H_3BO_3, and the specimens were electro-healed at 40 or 55C for 2h, using a current density of $4A/dm^2$. The electro-healed samples were then annealed (673K, 2h) in order to remove internal stresses. The artificial cracks were some 1.5mm in length, with central openings of 20 to 50µm. Crack-free control samples were prepared by using the same methods. In order to eliminate machining stresses, the samples were annealed (573K, 5h) at each stage. Specimens which were defined as cracked, uncracked or healed, with thicknesses of 100, 150, 200 and 300µm, were prepared for mechanical testing. Following electro-healing, the original crack was entirely filled with nickel crystals, with no visible voids on either side of the sample. Cross-sectional analysis of 100µm-thick samples revealed the presence of voids and unhealed zones near to the crack centre, and the unhealed volume increased with depth, attaining a few percent at about 50µm-deep. Micron-sized pores were often found in narrow crack regions, and no crystal growth was observed at tips with openings smaller than 1µm. High-resolution observation of the crack centre showed that nickel crystals nucleated on both crack surfaces and grew towards one another. They first formed ultra-fine equiaxed grains, followed by columnar growth. Growth twins were frequently observed in larger columnar grains. The healing crystals were very pure (circa 99.96t%). Careful analysis of the meeting point of converging crystal growth revealed the occurrence of atomic-level bonding between the crystals growing from opposite directions. The resultant interface resembled a typical grain boundary, similar to that between grains both growing in the same direction, thus confirming that conventional grain-boundary formation could occur at the crack-centre during a healing which was governed by electrocrystallization at the crystal/solution interface. A healing efficiency of some 96% was found for 100μm-thick samples. For 300μm-thick samples, it was about 28%. There was a good agreement between the measured and calculated strengths for various thicknesses. A decrease in the healing efficiency, with increasing thickness, was observed and amounted to about 33% at 300μm. This fall was attributed to an increase in unhealed regions, due to a non-uniform current density distribution and to early closure of crack openings.

Molecular dynamics simulations were used[53] to determine the fate of cracks within ultrafine-grained nanocrystalline (7nm) nickel during creep-deformation. The internal nanocracks were healed within a few picoseconds of the initial creep progress, even if the constant applied load on the specimen was tensile, and acted normal to the crack surface in the outward direction. This crack-healing behaviour was attributed to stress-driven grain-boundary migration, grain-boundary diffusion and the amorphization of specimens. The presence of nanocracks within ultrafine-grained nanocrystalline nickel even improved the creep properties slightly, and this improvement increased with increasing size of the internal cracks. The creep-curves of nanocrystalline nickel containing internal nanocracks with various sizes were compared with those of perfect nanocrystalline nickel under a constant load of 1GPa at 1309K. Creep in ultrafine-grained nickel was controlled mainly by grain-boundary diffusion, with the Coble mechanism being predominant. This rendered the creep deformation very rapid, and that was attributed to the fact that nanocrystalline metals exhibited a greater atomic diffusibility at the grain boundaries when compared with conventional polycrystalline materials. The creep-curves of nanocrystalline nickel with a pre-existing internal nanocrack were shifted slightly towards lower creep strains when compared with the creep-curves of perfect nanocrystalline nickel. This was attributed to the fact that the diffusivity was slightly lower for nanocrystalline nickel with an internal nanocrack when compared with crack-free material. The diffusivity, before and after healing the nanocrack, was determined in order to monitor the effect of crack-size upon the diffusivity during creep deformation as well as the diffusivity following complete amorphization. The diffusivity immediately following nanocrack-healing was slightly higher for specimens containing a large nanocrack than for specimens containing a small nanocrack, or no nanocrack. The diffusivity following complete amorphization was slightly lower for a specimen with a large nanocrack than for a specimen with a small nanocrack. The diffusivity increased markedly following amorphization, as compared with the stage just after nanocrack-healing. Appreciable variations in the grain structure indicated the occurrence of grain-boundary mobility due to stress. The generation and movement of crystal defects were observed. Some linear defects (disclinations) bound the crack surface to the main body of the specimen. Disclinations and stress-driven grain-boundary migration healed nanocracks in the nanocrystalline nickel. The grain-boundary region broadened as creep-deformation proceeded. The onset of grain-boundary widening was observed near to the nanocrack surface, unlike parts of the specimen. This was attributed to the congregation of atoms at the grain boundaries. Localized amorphization near to the nanocrack surface occurred at a very early stage and then amorphization of the entire specimen took place. Healing of an internal nanocrack occurred at the same time due to stress-induced grain-boundary migration and grain-boundary diffusion. The movement of atoms near to, or at, the nanocrack surface was traced. The nanocrack was healed by the movement of those atoms. Atoms at, or near to, the nanocrack surface and nearby grain boundaries moved towards the hollow region of the nanocrack region. This led to healing of the nanocrack with progressing creep deformation. Nanocrack-healing occurred slightly earlier in the case of nanocrystalline nickel with a larger crack as compared with a smaller crack. It was concluded that surface-diffusion as well as grain-boundary diffusion near to the nanocrack were critical factors for crack-healing during creep-deformation in the case of nanocrystalline nickel. The atomic diffusivity was greater in the region close to the nanocrack than in region farther from the nanocrack. This led to more rapid closing of the nanocrack during creep, and to faster amorphization in the region near to the nanocrack. A nanocrack was also equilibrated at 1309K for 210ps in order to determine the effect of temperature upon nanocrack-healing. Healing of the nanocrack did not occur in this case. A slight reduction in nanocrack-size near to 210ps was observed when the specimen was subjected to a zero load at 1309K. Creep-deformation simulations of specimens with a nanocrack were performed under a 0.1GPa load at 300 and 500K. The nanocrack healing-time was essentially equal at 300 and 500K for a given load. The applied stress was the predominant factor in crack-healing during creep, rather than the creep temperature.

Molecular dynamics and embedded-atom potential methods were applied[54] to pure nickel, and to nickel doped with rhenium, ruthenium, cobalt or tungsten at 300K. The object was to understand crack formation in the doped nickel in the (010)[001] orientation. When the nickel was doped with rhenium, ruthenium and tungsten, the matrix had increased lattice-trapping limits and this improved the mechanical properties. This in turn prevented bond-breaking at the crack-tips and promoted crack-healing. The average atomic and surface energies increased when rhenium, ruthenium and tungsten were added. Analysis of the energy increases clarified the effect which these additions had upon the lattice-trapping limits. The fracture strength of the nickel matrix at 300K increased due to the formation of stronger Ni-Re, Ni-Ru and Ni-W bonds. Doping the nickel with cobalt did not have any strengthening effect because Co-Ni bonds were weaker than Ni-Ni bonds. Doping with tungsten gave the best results.

Molecular dynamics simulations were used[55] to investigate the interaction between collision cascades and pre-existing nano-cracks in nickel, showing that collision cascades caused cracks to heal when such a cascade initiated a thermal spike, the core of which overlapped the crack. The crack-healing affected the distribution of radiation-induced defects which were generated during a collision-cascade. Defect-clusters such as dislocations and stacking-fault tetrahedra predominated in the presence of crack-healing, whereas isolated point defects predominated in its absence. Irradiation could thus reduce the number of pre-existing crack-like flaws in metallic components and thereby perhaps improve the mechanical behaviour.

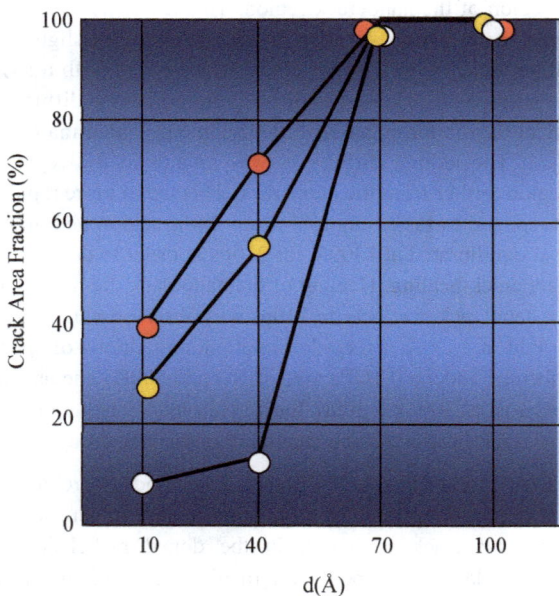

Figure 8. Remaining crack-area fraction in nickel as a function of PKA distance from the crack. Red: 0°, 2-layer crack, orange: 45°, 2-layer crack, white: 90°, 2-layer crack

The response of a 2-layer nanocrack to a primary knock-on atom initiated 10Å from the crack, perpendicular to the crack-surface and with a 10keV kinetic energy, was studied. The primary knock-on atom travelled towards the crack centre and displaced the atoms in its path from their perfect lattice sites. At 0.3ps after primary knock-on atom initiation, collisions between the primary knock-on atom and other atoms had initiated a thermal spike that overlapped the nanocrack. Local melting and atomic mixing occurred in the thermal spike core. Following the ballistic phase of the collision cascade, the material within the thermal spike cooled and crystallized, leaving radiation-induced defects. At 26.1ps after primary knock-on atom initiation, the model simulation contained stacking faults isolated vacancies. The nanocrack was no longer visible, indicating that parts of it had been closed during the collision cascade. An examination was made of the final crack structure after the collision-cascade, viewing it along the crack-surface normal, <111>. Most of the crack had been closed, while other parts remained open. The fraction of original crack surface atoms that remained after the collision-cascade was some 8.5%. The collision-cascade thus caused most of the crack area to close.

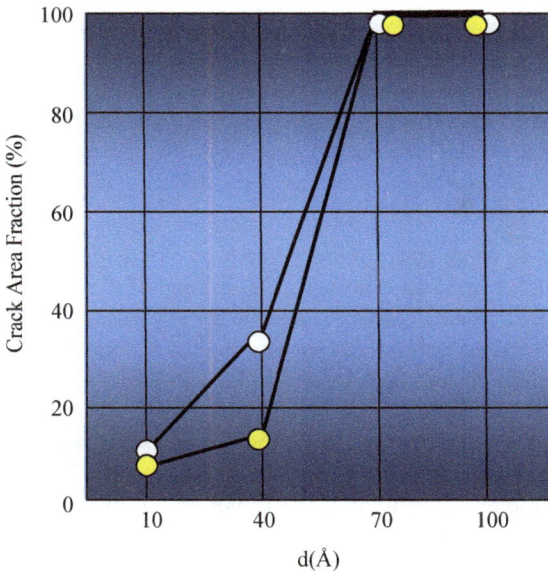

Figure 9. Remaining crack-area fraction in nickel as a function of PKA distance from the crack. White: 90°, 2-layer crack, yellow: 90°, 3-layer crack

Cascade-induced crack closure occurred only when the thermal spike overlapped with the crack. A further study was made of a collision-cascade which was caused by a 10keV primary knock-on atom 70Å from the crack, perpendicular to the crack surface. Even at its maximum extent, the thermal spike core did not overlap with the crack. Following the ballistic phase, isolated point defects alone survived. No dislocation loops nor stacking faults were observed, and the nanocrack remained unaffected. Thermal spikes emit compressive stress-waves that can affect the response of a surrounding solid. Inspection of the present collision-cascade showed that any stress-wave which it emitted then interacted with the crack. This caused transient closing of the crack surfaces, but not enough to heal the crack. Crack-closure was associated with the formation of stacking faults, and there were no stacking faults when there was no crack-closure. It was suggested that cascade-induced crack-closure involved conversion of the initial nanocrack into a vacancy-rich dislocation network, including stacking-fault tetrahedra. In the case of crack-closure, the remaining dislocation network was far more extensive than when no crack-closure occurred. The defect structure in the latter case was dominated by isolated point defects. For a given primary knock-on atom direction, the fraction of crack-surface atoms increased with increasing initial primary knock-on atom distance from the crack. The greater the primary knock-on atom distance from the crack, the less likely it was that such overlap would occur and therefore less likely the occurrence of crack closure. It was noted that primary knock-on atoms with velocities which were most nearly perpendicular to the crack surface were more effective in closing the crack. This was attributed to the larger fraction of crack-area which overlapped with the thermal-spike core.

No crack-closure was observed for primary knock-on atoms which were more than 70Å from the crack. This was because there was no overlap of the resultant thermal spikes with the crack. The remaining crack-area fraction for 90° collision cascades and 2- or 3-layer cracks was such that thicker crack underwent less healing. No healing occurred in either case when the primary knock-on atom distance from the crack was greater than 70Å, because of the absence of overlap between the thermal spike and the crack. Primary knock-on atoms which were less than 70Å from the crack led to a high dislocation-line length. This was consistent with the formation of vacancy-rich dislocation networks due to crack-healing. The dislocation-line length at 90° was greater than at 45° and in turn was greater than that at 0°.

This was again consistent with a greater thermal-spike overlap with the crack for primary knock-on atoms which were oriented normal to the crack. For a primary knock-on atom distance of 70Å or greater, the dislocation-line length was comparable regardless of the primary knock-on atom velocity and crack thickness. Almost no dislocations were formed because no crack-healing took place. The number of point defects in the absence of crack-healing was 3 times the number of point defects when crack-healing occurred. Crack-healing thus had a marked effect upon the types of radiation-induced defects which were generated by collision-cascades. The fraction of crack-area which remained was determined as a function of primary knock-on atom kinetic energy for events which were initiated at a distance of 100Å from the crack and for velocities perpendicular to the crack surface (figures 8 to 10). As compared with the 10keV case, the remaining crack-area fraction for 20 and 30keV primary knock-on atoms was much smaller; a reduction from almost 100% to some 23%. The difference was attributed to the larger thermal spikes which were produced by primary knock-on atoms of greater energy. Larger thermal spikes could overlap with, and heal, cracks which were further from the initial primary knock-on atom.

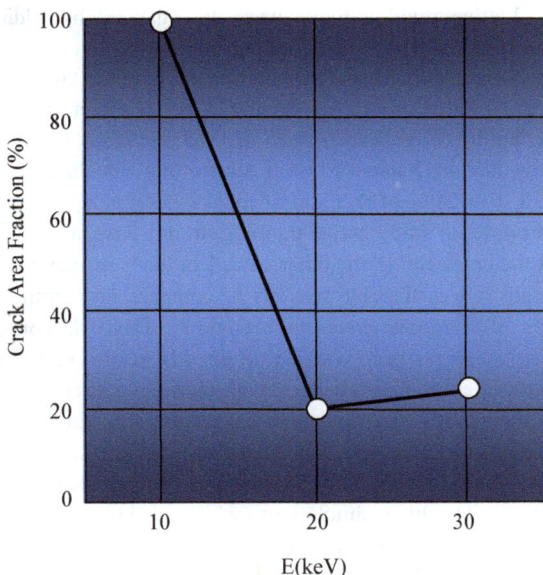

Figure 10. Fraction of crack-area remaining in nickel after collision cascades as a function of the PKA kinetic energy

The thermal spike was a volume of material within which the deposited energy raised the temperature to thousands of degrees; the material melted and remained so for 1 to 2ps. When the thermal spike overlapped with a crack, the liquid could enter the crack by diffusion or pressure-driven flow. In the case of diffusion the maximum distance travelled was proportional to the product of the square-root of the liquid diffusivity and the spike-duration; a maximum distance of only some 1.5Å. It was concluded that pressure-driven fluid flow was the more likely healing mechanism. On the other hand, the observed dislocation networks could form in the solidifying core of the thermal spike.

Ni₃Al

The effect of rhenium and tungsten upon the brittle fracture of Ni_3Al cracks with 3 orientations was investigated[56] via molecular dynamics, using embedded-atom method potentials and discrete variational methods for Ni-Al-Re and Ni-Al-W systems,. It was found that rhenium and tungsten could strengthen Ni_3Al alloy and improve the deformation resistance. Lattice-trapping limits were not changed by adding a single rhenium or tungsten atom, but increased upon adding 1 or 2at% of rhenium or tungsten atoms. The rhenium and tungsten could prevent bond-breaking at the crack-tip and could promote crack-healing in Ni_3Al, with tungsten having a greater effect than did rhenium. In order to judge the reliability of embedded-atom method potentials for treating Ni-Al-Re and Ni-Al-W systems, key mechanical properties were first calculated for defect-free Ni_3Al which contained 0, 1 or 2at% of rhenium or tungsten. An analysis was made of 2 doping configurations in order to study lattice-trapping. In one case, a single rhenium or tungsten was located at the crack-tip. In the other case, 1 or 2at% of rhenium or tungsten was randomly distributed. It was thereby possible to compare how both isolated and collective dopant atoms affected the crack-tip behaviour of Ni_3Al. For each crack-orientation, a single rhenium or tungsten was placed directly at the crack-tip. The bond lengths along the (010)[001] crack-front changed with increasing stress-intensity. Small distinct jumps in bond-length revealed an upper lattice-trapping limit for non-doped Ni_3Al. For both rhenium and tungsten single-atom doping, the critical bond-length jump occurred within the same stress-range. This implied that neither element affected the upper or lower trapping-limits. Other simulations, which placed a single rhenium or tungsten atom at alternating positions along the crack-front, confirmed those results. Neither the trapping-limits nor the trapping-ranges varied as a function of atom-placement. Bond-breaking occurred simultaneously across multiple atomic rows at the crack tip. This confirmed that fracture involved a broad atomic front. A minor bending of atomic rows around the Re/W atoms before fracture implied that there were stronger Ni-Re and Ni-W interactions as compared with that of Ni-Al. In spite of this local bending, Ni-Re and Ni-W bonds broke at the same time as Ni-Al bonds. This showed that the local distortions did not alter the overall lattice-trapping characteristics in the single-impurity case. Four different random distributions of 2at% of rhenium and tungsten atoms were simulated for the (010)[001] crack-orientation. There were minimal variations in critical values across these configurations and this validated the use of random doping. Both lattice-trapping limits and trapping ranges increased at higher dopant levels. In models which were doped with 2at% of rhenium, critical values increased for all crack orientations, as compared with non-doped Ni_3Al. This suggested that rhenium and tungsten improved crack-resistance by suppressing bond-breaking at the crack-tip. The tungsten had a greater effect than rhenium and this was attributed to a stronger bonding between nickel and tungsten than between nickel and rhenium. In spite of the improvements in crack-resistance, the overall lattice-trapping ranges remained narrow for all dopant-levels and orientations. Lattice-trapping varied with crack orientation. This again support the idea that cleavage-fracture in Ni_3Al is directionally anisotropic.

Ni-Al-Mo

A study was made of the effect of selective laser-melting upon the tendency to defect-formation in Ni-5.4Al-3.8Mo-2.5wt%Ta alloy (VZhL21)[57]. It was possible to construct a processing window which identified the optimum processing parameters for producing a structure which contained a minimum number of pores and cracks. A 2-stage treatment was proposed for healing cracks. The first stage was vacuum heat treatment for partial healing of cracks, and the second stage was hot isostatic compaction for healing of the remaining defects. In order to start the diffusion-healing of cracks, the temperature had to be close to that for complete γ'-phase dissolution. The holding-time was also important; with a short (<4h) exposure time, insufficient diffusion occurred and the efficiency of the treatment was low. An exposure time of at least 4h was required. The diffusion-healing process for cracks ceased when the crack-energy reached a minimum and it transformed into a defect that resembled an extended pore.

Ni-Cr

The solidification behaviour and weld solidification-cracking susceptibility of high-chromium nickel-based filler metals were investigated[58]. Two heats of ERNiCrFe-13 (with 52MSS filler), one heat of ERNiCrFe-7A (with 52M filler) and a single heat of modified ERNiCr-3 (with 52i filler) were chosen. The 52i filler had the widest solidification range, followed by the two heats of 52MSS and 52M. Filler 52i also had the widest eutectic temperature range. The interdendritic eutectic which formed in the weld metal of this filler, and of 52MSS, was enriched in niobium, which resulted from the eutectic reaction, of $\gamma + l \rightarrow \gamma + NbC$, at the end of solidification. Both heats of 52MSS, and of 52i, were more susceptible to solidification-cracking than was 52M. A slightly higher resistance to solidification-cracking in 52i, relative to that of 52MSS, was attributed to crack-healing during the final stages of solidification. This was due to the higher fraction of eutectic liquid in 52i. That is, the slightly better resistance to solidification-cracking at lower strain-levels in filler-metal, 52i, as compared to 52MSS-A and 52MSS-B, could be explained by the effect of crack-healing provided by the larger fraction of eutectic that is formed in this weld-metal during the final stages of solidification. Examples of back-filling along the solidification grain boundaries in 52i, 52MSS and 52M were observed. The region of back-filling could be detected by a change in the solidification-structure across the solidification grain-boundary. A relatively wide back-filled region existed at the crack-tip in 52i and 52MSS weld-metal, but less back-filling was observed at the crack-tip of 52M. A higher level of the eutectic constituent in the 52MSS and MLTS-2 weld-metals, than in 52M, was predicted, reflecting the higher niobium content of these filler-metals.

Ni-Cr-Fe

Niobium and molybdenum were added to Ni-29Cr-9wt%Fe (690) base alloy in order to investigate the effect of eutectic formation at the end of solidification with regard to the back-filling and healing of solidification cracks[59]. Solidification-cracking was introduced by using cast pin tear testing. Regions of back-filling were identified and were characterized by using optical and electron microscopy. The degree of crack back-filling markedly increased with increasing niobium content, while the addition of molybdenum did not appear to affect the amount of eutectic nor the degree of back-filling. The eutectic composition was constant, and increasing the niobium content to above 4wt% had little effect upon widening the solidification range. It had a beneficial effect upon reducing solidification cracking, as a result of a crack-healing effect. In detail, niobium could lead to an increased susceptibility to solidification-cracking because its segregation promoted a terminal eutectic reaction and a widening of the solidification temperature-range. High fractions of eutectic which form at the end of solidification are associated with a form of crack-healing which occurs when cracks are back-filled by the liquid eutectic, if the amount available is sufficient. The addition of up to 4wt%Nb increased the cracking susceptibility but, at higher levels, there was a decrease which was attributed to back-filling. Back-filled regions were sparse and could be disconnected in 2wt%Nb samples, possibly due to the lower fraction of eutectic present. The 6wt%Nb samples featured a more extensive and continuous network. There seemed to be no effect of molybdenum at either niobium level although, in 6wt%Nb, back-filled cracks appeared to be wider. It was unclear whether this was a truly niobium-related effect. It was difficult to determine the effect of molybdenum, although some reduction in cracking occurred when molybdenum was added to 6wt%Nb alloys. There was a clear difference in the effects of eutectic fraction upon back-filling when comparing low-Nb and high-Nb alloys. A 2wt%Nb sample exhibited a completely healed crack along a solidification grain-boundary, even though the eutectic fraction was relatively low. An 8wt%Nb alloy exhibited an extensive back-filling which had almost completely healed a crack. Only a small amount of open crack remained. The niobium was concluded to exert a predominant effect upon the formation of eutectic liquid and therefore upon the degree of back-filling. Because molybdenum additions had no apparent effect upon the fraction of eutectic available during solidification, it was suggested[60,61] that molybdenum might increase wettability of the eutectic liquid for grain boundaries and permit crack back-filling to be more efficient. Molybdenum additions might therefore permit more complete crack-healing and thus a further reduction in cracking susceptibility when the fraction of eutectic was sufficiently high. Steps were taken to induce local re-melting and wetting of eutectic in weld-metal samples, with and without molybdenum additions. The results suggested that molybdenum somewhat affected the eutectic re-melting and/or wetting characteristics.

Ni-Cr-Co

A considerable disadvantage of the IN738LC superalloy is its tendency to crack during laser powder-bed fusion. The effect of a 400W laser and 90° rotation upon crack formation, and the ability of hot isostatic pressing post-treatment to reduce them, was investigated[62]. There was a limiting crack-width of 6μm, beyond which hot isostatic pressing was unable to achieve crack-healing. The treatment meanwhile led to microstructural changes, with massive precipitation of γ'-phase. Its formation was associated with an increase in microhardness of up to 23%. The crack-density was measured in hot-isostatically pressed samples and compared with that in as-prepared samples. The mechanism of crack-closure during hot isostatic pressing involved 2 principal phenomena: the stress-gradient around the cavity and atomic diffusion. The most efficient sinks for vacancies were grain boundaries, and vacancies travelled from the crack surface to grain-boundaries via grain-boundary diffusion while atoms diffused in the opposite direction. All of the cracks were located at grain boundaries, and this healing process was therefore favourable, but hydrostatic pressure was also required in order to promote atomic diffusion and increase crack-healing. The choice of an adequate treatment-temperature was important, and that was found to be 1180C; higher than the solvus temperature of the γ' phase. It had been proposed that crack-elimination during hot isostatic pressing was related to the average crack-width, but the present results proved that crack-length was not the most important factor in the present case, where all of the cracks were shorter than 300μm and not all were eliminated. The atomic diffusion distance was limited, and this made it impossible to remove all the cracks.

A liquid-induced healing post-processing treatment was studied for the correction of defects that were produced by laser powder-bed fusion additive manufacturing. It enabled complete healing of the micro-cracks by inducing a solid-liquid phase transition in the cracked regions, thereby improving the mechanical properties. The IN738LC alloy, prepared using laser powder-bed fusion, was employed[63] as a test-bed for investigating the effects of a re-melted liquid fraction and isostatic pressure during crack-healing. When compared with existing methods for crack-elimination, such as narrowing the solidification range, promoting equiaxed grain growth, reducing thermal stress and hot isostatic pressing, the new method led to better crack-healing. The liquid-induced healing process permitted the healing of cracks in printed metallic components. A crack usually occurred in the grain-boundary region, where it solidified last and melted first when re-melted. The healing process involved 3 steps. The first step was to heat the component into the brittle temperature range under vacuum so that micro-remelting occurred at the grain boundaries and re-melted liquid back-filled the cracks. In step-two, when the material had melted by less than 10vol% back to roughly where cracks began to initiate during laser powder-bed fusion, isostatic pressure of less than 5MPa was applied. In step-three the component was cooled at a controlled rate under the isostatic pressing in order to prevent secondary cracking and shrinkage porosity. The window for the healing process was identified by considering the re-melted liquid fraction which was the control-mechanism for the re-melting temperature, and the isostatic pressure. Being a typical non-weldable alloy, IN738LC was a very good test-subject for the present healing approach. The grain boundaries gradually melted as the specimen was re-heated to above the solidus temperature, thus forming an interconnected liquid film which back-filled the cracks. The specimen was in a dense semi-solid state at that stage. The liquid film then solidified during subsequent cooling, demonstrating the efficiency of welding of the cracks at the micron scale. The process was applied to as-printed IN738LC with an initial crack fraction of 0.89vol%. The sample was re-heated to 1290C in order to produce intergranular liquid amounting to about 9.1vol%. The vacuum environment at that stage ensured the back-filling of surface-connected cracks. Compressed argon was then introduced into the system so as to impose an isostatic pressure of 1MPa. The material was finally cooled at 2C/min until complete solidification occurred. Calculations were made of the volume and sphericity of the porosity defects, and it was conjectured that the porosity in as-printed IN738LC existed mainly in the form of cracks because of the high volume and low sphericity. All the cracks, even the largest, were healed with a volume of $1.66 \times 10^6 \mu m^3$. A few micropores alone remained following the treatment. The evolution of the spatial distribution of cracks, and the increase in relative density of specimens after the treatment, indicated that the simple approach could be very effective in healing microcracks, including the surface-connected ones. Control of the volume fraction of liquid was a critical factor. Increasing the heating-rate slightly accelerated the melting process. In order to control closely the liquid fraction during the process, the heating-rate for governing the relationship between liquid-fraction and temperature had to be consistent. The micro re-melting interval for healing was selected according to the alloy's brittle temperature range so as to ensure that the material was partially re-melted back to the point where solidification-cracks began to nucleate during laser powder-bed fusion. The liquid-induced healing process was also applied without imposing any isostatic pressure in order to identify the optimum re-melting temperature which corresponded to the initial degree of cracking. Assuming that the crack was back-filled via volume expansion of the re-melted liquid, the volume expansion which was caused by the re-melting was expected to be the same as the initial crack-volume when choosing the optimum temperature to ensure that the volume of the component did not expand due to that re-melting. The relationship between the crack-volume and optimum re-melting temperature could be derived. A comparison was made of theoretical predictions and experimental results on optimum temperatures for various degrees of initial cracking (figures 11 and 12). This confirmed that volume expansion of the re-melted liquid promoted back-filling of cracks. In the event of insufficient re-melting, the volume expansion which was caused by the re-melted liquid was insufficient to back-fill the cracks fully, leaving residual cracks. On the other hand, excessive re-melting led to complete back-filling but also engendered a volume expansion which increased the risk of shrinkage porosity occurring during subsequent solidification. The application of isostatic pressure was another critical factor for the success of the treatment. Because re-melting occurred, the appearance of shrinkage porosity was inevitable during subsequent solidification if no feeding was possible. In order to counter this problem, isostatic pressure was applied to the material at the micro-remelting stage. This material exhibited an elastic-viscoplastic response under compressive stress and was expected to be volumetrically compressed in order to compensate solidification shrinkage. Uniaxial compression tests of the micro-remelted alloy IN738LC could approximately determine the stress-strain responses. The compression curves revealed elastic variation for strains ranging up to about 1%, and exhibited anisotropy with increasing strain. The minimum pressure required for liquid-induced healing depended upon whether the corresponding strain exceeded the solidification shrinkage. In order to imitate deformation under isostatic pressure during the process, compression at a constant stress of 1MPa was performed under continuous cooling. The strain was greater than the linear solidification shrinkage from 1290C, for a liquid fraction of 9.1vol%. This demonstrated that the strain under a stress of 1MPa was sufficient to counteract the solidification shrinkage and thus

Materials Research Forum LLC
https://doi.org/10.21741/9781644903773

prevent possible porosity. As-printed material with an initial crack of 0.889vol% was subjected to liquid-induced healing, using an isostatic pressure of 1MPa. All of the cracks were healed without introducing shrinkage porosity. When the treatment was repeated, but without applying an isostatic pressure, some 0.3vol% of shrinkage porosity remained, for the same degree of initial cracking. The underlying mechanisms of linear-induced healing and hot isostatic pressing were distinct, in spite of their both exploiting isostatic pressure at high temperatures. The crack-closure mechanism of hot isostatic pressing relied upon plastic deformation at the micron scale and upon subsequent diffusion. It therefore required extreme pressures and lengthy soaking, and was incapable of healing surface-connected defects.

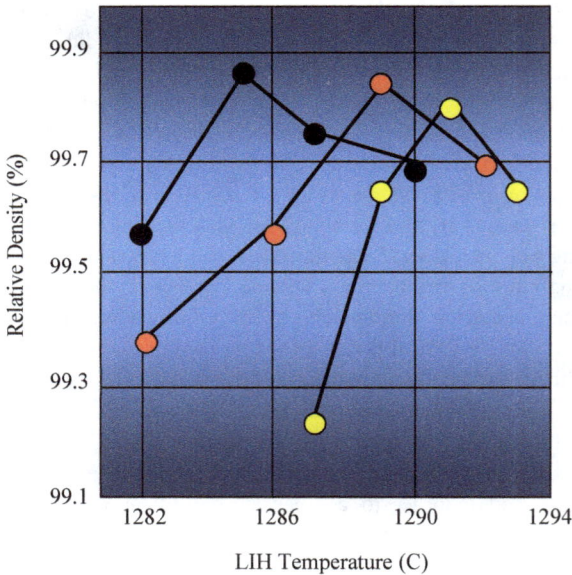

Figure 11. Relative density after liquid-induced healing of IN738LC as a function of the initial crack volume fraction (%). Black: 0.868, red: 1.276, yellow: 1.427

The liquid-induced crack-healing mechanism instead involved the instantaneous back-filling of intergranular liquid, with the isostatic pressure preventing potential porosity appearance during solidification of the liquid film. The required pressure was minimal, and surface-connected defects could be effectively healed. In the case of a sample which contained some 1.4vol% of initial cracks, only those which were independent of the surface were closed by hot isostatic pressure. The latter could however lead to crack-expansion, due to the extremely high pressure within the cracks. The sample was then subjected to liquid-induced healing, following the pressing. Residual cracks were largely eliminated by applying this approach. The microstructural evolution during liquid-induced healing was also investigated. In addition to healing of the cracks, columnar grains underwent a marked equiaxed transformation and slight coarsening. The fraction of high-angle grain boundaries increased and the crystallographic orientation changed. The isostatic pressure which was required for liquid-induced healing was usually less than 5% of that required for hot isostatic pressing.

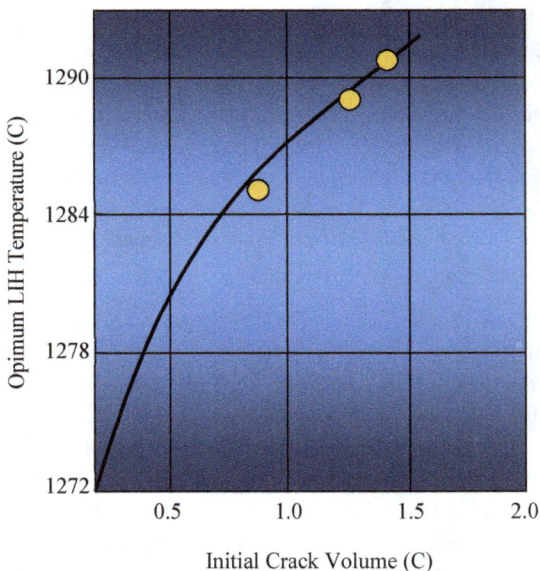

Figure 12. Optimum temperature for liquid-induced healing of IN738LC as a function of the initial crack volume fraction

A new heat-treatment approach which involved pre-softening before hot isostatic pressing was developed[64] in order to drive crack-healing in René 142 (Ni-18Cr-10Co-

9Mo-3Ti-1.4wt%Al) superalloy which was prepared by laser powder-bed fusion. This alloy exhibits a tendency to severe cracking for a wide range of printing parameters. This leads to the formation of solidification and liquation cracks. The cracking was due mainly to the wide solidification range, to the presence of a liquid film and to the concentration of residual stresses. The use of pre-softening solution heat-treatment markedly reduced the dislocation density and residual-stress levels. Subsequent hot isostatic pressing produced a defect-free dense structure having a yield strength of 850MPa, an ultimate tensile strength of 1227MPa and an elongation of 13.7%. Hot isostatic pressing was generally used for the post-processing and defect-healing of the products of additive manufacturing. The effectiveness of the process depended upon the temperature and pressure. Alloys possessing high-temperature strength required higher temperatures and pressures. The properties René 142 alloy necessitated the use of 1200C and 140MPa. During direct hot isostatic pressing of the alloy, most cracks persisted. When a pre-treatment (1260C, 1h; 1275C, 1h; air-cool) was added, there was then marked crack-healing, leaving only a few voids. Direct hot isostatic pressing alone failed to heal cracks and also promoted crack propagation due to high residual stresses and their release. The porosity of laser power-bed fused alloy first decreased and then increased with increasing printing volumetric energy-density. The main faults were solidification and liquation cracks, distributed mainly along high-angle grain boundaries. Internal micro-cracks and other defects could be eliminated by the combined treatment, leaving defect-free dense material with pseudo-equiaxed grains (120 to 150μm) and square 540nm γ'-precipitates.

Ni-Cr-Nb

The feasibility of using electric pulses to heal surface cracks on the nickel-based alloy, GH4169 (Inconel 718), under compressive loads was investigated[65]. There was an incubation time for the onset of crack-healing, and this delay was associated with the local temperature at the crack tip. The crack-size decreased with increasing pulsing time, for a given constant healing-rate before complete healing of the crack occurred. Increasing the compressive load accelerated the healing process. Electric pulsing led to the formation of a zone of influence which surrounded the crack, with the finest grains being found in the healed crack and the coarsest grains being found away from the zone. The indentation hardness increased with increasing distance to the tip of the healed crack. A model of viscous flow in the crack channel was used to describe the crack-healing. The resultant force on the crack faces, due to crack-healing, increased with increasing healing time and with decreasing crack-width. The cracks experienced various durations of electric pulsing under a compressive load of 500N, and the crack-length decreased with increasing pulsing-time. With a pulsing-time of 3.5s, the crack was almost entirely healed. In the absence of a compressive load, there was essentially no healing of the crack. An increase in the compressive load to 700N led to complete healing of the crack. Local discharges occurred under a compressive load of 1000N, leading to opening of the crack. The local discharge (explosion) was attributed to a short-circuit which caused ionization of the air. Aluminium, oxygen, molybdenum, iron, cobalt, titanium, niobium and chromium (constituents of the alloy) were detected around a healed crack, confirming the occurrence of mass-transport to the crack during pulsing. Markedly high titanium and aluminium segregation at the healed crack was attributed to the high cooling-rate during the pulsing. The rapid cooling-rate of solidification impeded the diffusion of titanium and aluminium from the molten zone. The slight segregation of oxygen to the healed crack was attributed to adsorption from the air. Fine grains formed near to, and at, the healed crack and delineated the zone-of-influence of the electric pulse for healing the crack. The grain-sizes ranged from 2.0 to 37.2μm, with an average size of 16.6μm. The grain sizes remote from the zone-of-influence ranged from 64.5 to 1546.4μm, with an average size of 930.0μm. The large difference in the average grain sizes was attributed to rapid cooling following pulsing, and suggested that the pulsing caused local melting around the crack tip and to healing of the crack via the growth of new grains under compression. The variation of indentation hardness with distance to the tip of the healed crack was such that the hardness was lowest in the healed zone and highest in the base metal. This trend was attributed to the presence of the newly-formed grains in the healed zone, where the dislocation-density was much lower than in the base metal. The evolution of the crack-length under a compressive load of 500N was such that the temperature near to the crack increased rapidly at the onset of electric pulsing and slowly approached a plateau as the pulsing time reached 2s. The highest temperature attained was 1140C; slightly less than the melting-point (1260-1340C) of the alloy. The measured temperature was not expected to reflect that at the crack tip, and to be higher than 1140C to the extent that local melting occurred at the crack tip. The onset of crack-healing in the alloy under a compressive force of 500N began at a pulsing duration of 1.5s and healing was complete after a pulsing duration of 3s. There was a regime within which the crack-length decreased linearly with pulsing-time between 1.5 and 2.5s (figure 13). For pulsing-times ranging from 1.5 to 3s, it could be assumed that the temperature at the crack tip remained almost constant. Healing of the crack could therefore be treated as being an isothermal process. The healing-rate of the crack was deduced to be about 2.94mm/s. In general, many factors contributed to the variation in indentation hardness. This included the dislocation-density and grain-size; the smaller the grain-size, the higher the hardness. The higher dislocation density, the higher the indentation hardness, with individual contributions being made by grain-boundaries and essentially dislocation-free grain interiors. So, in summary, electric pulses of 2.511V with a pulse-width of 25ms and a frequency of 20Hz, under a compressive load of 500N, could heal 3.14mm-long surface cracks within 3.5s.

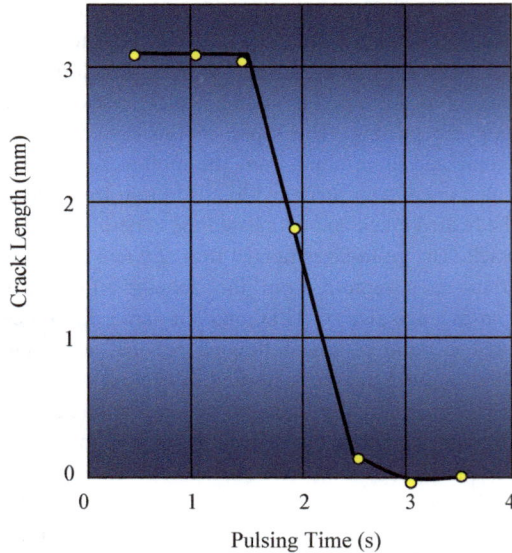

Figure 13. Temporal evolution of the crack-length during electro-pulse healing of GH4169 (Inconel 718) under a compressive force of 500N

Ni-Cr-W

A new *in situ* re-melting process was used[66] to produce almost crack-free laser powder-bed fusion-prepared Haynes 230 (Ni-22.2Cr-13.5W-2.2Mo-1.1wt%Co) alloy. Hot-cracking occurred mainly in the centre of deep and large molten pools. By limiting the re-melting depth to between 70 and 80% of the original layer-depth, most of the original cracks were re-melted. The remaining cracks were back-filled with liquid metal in the melt pool, and the cracks were healed. This largely avoided tungsten, chromium, molybdenum and oxygen segregation at grain boundaries and promoted the precipitation of high melting-point $M_{23}C_6$. The re-melting also melted obliquely growing dendrites on the sides, and coarse dendrites in the middle and top of the molten pool. This led to a reduction in crack-sensitive grain boundaries and grain-boundary serration, and relieved the increased residual stresses caused by re-melting. The higher dislocation density and more random grain orientations of re-melted specimens led to an increased contribution of dislocation-strengthening and to great plasticity. The re-melting had an appreciable effect upon crack inhibition, including the healing of cracks during laser action and the influence of the re-melted microstructure upon crack-resistance. During re-melting, most of the cracked area was melted and thermal cycling could re-melt fractured low melting-point phases on each side of the residual original cracked grain boundaries. Because cracking occurred mainly in the middle of the original molten pool, most of the melted grain boundaries of the remaining original molten pool belonged to the crack-emanating zone of original layers. The crack depths and widths were small, the liquid-film area was small, and the area where voids appeared during the second non-equilibrium solidification was greatly reduced. Even when there was crack extension, the shorter liquid-film length during subsequent solidification reduced the driving force for crack extension. In order to identify the crack-healing mechanism, crack-inhibition by various re-melting depths was studied. The smaller the re-melting laser-power, the smaller were the cracks. Low-energy re-melting reduced residual stresses more than did high-energy re-melting. It was suggested that the increase in crack-density resulted from an increase in crack back-fill resistance. The re-melting process permitted the melting depth of the original layer to reach 70 to 80%. This promoted liquid-metal back-filling of the cracks and led to crack-healing. The melting reduced the crack-density by more than 80% and permitted the preparation of nearly crack-free specimens.

Ni-Fe

The effect of 2-mercapto-5-benzimidazole sulfonic acid in a Watts bath upon crack electro-healing was investigated[67] by means of potentiostatic and galvanostatic voltammetry. The morphology of healed cracks indicated that the healing crystals grew in a controllable manner under forced convection upon adding the organic compound, and the healing crystals preferentially filled the crack tip up to the crack side-walls. No defects were observed along the interface between the substrate and the healing crystals, and this was attributed the much higher current efficiency provided by the organic. In detail, 2-mercapto-5-benzimidazolesulfonic acid has an acid-base nature which can be protonated in Watts solution. Studies of nickel, copper, Ni-Fe and Co-Fe alloy had shown that, during their deposition, the acid was adsorbed and formed a layer which inhibited the reduction of the metallic ions, leading to void-free trench-filling. This was characterized by preferential deposition at the bottom and by subsequent geometrical levelling. There was therefore a similarity between the method of electro-healing cracks and the filling of trenches, although the lengths of cracks were usually of the order of millimetres and had an opening of a few tens of micrometres. Cracks having such features offer quite different electrochemical conditions of additive transfer, distribution and consumption. The performance of the acid during the controllable electrohealing of a crack in nickel was determined. A rotating platinum disc electrode, rotating at 50 to 2000rpm, was used to model the effect of convection upon voltammetry of the acid-added electrolyte and simulate the electrochemical behaviour between crack-tip and crack-centre. In order to clarify the correlation between the electrochemical behaviour of additives, and the morphology of healed cracks with additives, data on polyethyleneimine use were compared. The healed morphology, following the use of 50µM and 10mA/cm^2 for crack electro-healing under forced convection, was inspected. This showed that the healing crystals grew preferentially from the crack-tip to the crack side-walls. This was attributed to the difference in inhibitor distribution at the tip and side-walls. The growth-modes operating during polyethyleneimine and 2-mercapto-5-benzimidazolesulfonic acid were similar, but there were marked differences between the morphologies produced by those inhibitors. There was a line-like void defect along the interface between the healing crystals and substrates when using polyethyleneimine. This defect was not present when the acid was used. Defects which resembled the annual growth-rings of trees were observed in both cases, although the rings were more densely-spaced when using polyethyleneimine. Light regions (ring-width) and dark regions (ring) alternated. In the light regions, the growth of the healing crystals could be traced. The widths of the annual rings after polyethyleneimine use were less than 0.5µm, while those following acid use ranged from a few micrometres to 30µm. The annual-ring pattern was suggested to be common when controlling the electrohealing of cracks by using an inhibitor. The light regions between rings were attributed to the active-growth mode of healing crystals, while the dark rings were attributed to passive inhibition and consequently negligible nickel deposition. The annual-ring pattern thus reflected a transition, of the growing front of healing crystals, from active to passive and back to passive when the acid-concentration was high and the degree inhibition was sufficiently high. Formation of these defects seemed to be unavoidable during controlled crack electrohealing when inhibitors were used. The inhibitors typically exhibited a concentration-dependent or convection-dependent effect, thus making it possible to achieve differential inhibition between crack tips and side-walls. At side-walls, where the mass-transfer of additives was favourable, a complete inhibition-state was produced. The healing crystals grew in a conformal growth-mode at an extremely low rate. At crack-tips, the mass-transfer of additives was unfavourable and the growth of healing crystals from the crack-tip was much less inhibited and exhibited an active growth stage. During this period, additives progressively accumulated via mass transfer. When the surface-coverage of inhibitors exceeded a critical level, complete inhibition occurred and the healing crystals exhibited passive growth, with an annual ring being formed. Following the passive growth stage, the surface-coverage of inhibitors was reduced due to inhibitor-removal or interface pH-shifts. When inhibition ceased, the growth of healing crystals returned to the active stage. The so-called annual ring was thus a transition between active growth and passive inhibition, with rings corresponding to complete inhibition and the intervals between rings reflected the accumulation of inhibitors. Predominant amounts of nickel, oxygen and sulphur were detected in the annual rings and were ascribed to 2-mercapto-5-benzimidazolesulfonic acid. In the active-growth stage between rings, only nickel was found, with no detectable oxygen or sulphur. When using a single inhibitor or inhibitors, the growth-mode of the healing crystals appeared to be controlled by the degree of inhibition. When this was low, as in the case of polyethyleneimine without agitation, the healing crystals did not exhibit a controllable growth pattern but rather the general growth-mode observed without inhibitor. When the degree of inhibition is high, as in the case of poyethyleneimine under forced convection, a controllable growth mode is attained but both interface defects and an annual-ring pattern was produced on the healed crack. The addition of 2-mercapto-5-benzimidazolesulfonic acid led to an improved cathode current-efficiency and thereby eliminated interface defects. In spite of the differences in the effects of 2-mercapto-5-benzimidazolesulfonic acid and polyethyleneimine, the annual-ring pattern was still observed upon adding the former. The overall effect of adding an accelerator was to de-polarize the reduction of nickel ions by which the growth of healing crystals from the side-walls would be enhanced. Growth from the crack-tips would meanwhile be less rapid, considering the convection-dependent distribution of accelerator. The de-polarization effect of the accelerator weakened the inhibition in the crack-tip and could eliminate the formation of annual ring-like defects.

Ni-W-Co

A study was made of the effects of hot isostatic pressing upon the crack-healing and mechanical properties of laser repair-welded CM247LC (Ni-9.5W-9.2Co-8.1Cr-5.6Al-3.2Ta-1.4wt%Hf) precipitation-hardened superalloy[68]. The method involved direct re-melting, and simulated repair welding with a filler which had a chemical composition that matched that of the base superalloy. Various types of crack, including solidification cracks in the weld fusion-zone and divers types of liquidation cracks in the heat-affected zone, were observed. By means of proper hot isostatic pressing healing process, all of the cracks in the weld fusion zone and the heat-affected zone of the repair-welded pieces were healed. Nano- and micro-sized carbides tended to form discontinuously along the trace of the healed crack. Previous studies had shown that hot isostatic pressing could eliminate voids and pores from a casting, and heal cracks in laser-deposited material. In order to test the feasibility of using hot-isostatic pressing for crack-healing in homogeneous filler-wire repair welding, a study was made of the direct laser surface re-melting of CM247LC. When cracks are present, an IN-625 (Ni-23Cr-10Mo-4.15Nb-1wt%Cr) filler wire is employed for surface sealing before hot-isostatic pressing. A longitudinal repair-weld which was about 10mm long and about 0.5mm wide was generated. The inside of a surface crack could contain gas or impurities that impaired healing. The hot isostatic pressing involved applying 175MPa for 3h at 1180C. A small part of the IN-625 and the original CM247LC was mixed and no surface cracks were found on top of the weld fusion-zone. At this stage, the original CM247LC solidification crack at the bottom of the weld fusion zone and a liquation crack in the heat-affected zone both still existed. After 3h at high temperature and high pressure, solidification and liquation cracks had completely healed, various zones existed in the weld fusion-zone of repaired welded samples: a top weld fusion-zone, a partial mixed zone, the weld fusion-zone, an un-mixed zone and a bottom weld fusion-zone. The unmixed zone was a peculiarly-structured zone which resulted from laser repair-welding. Due to the low energy and very rapid cooling-rate of the laser process, the IN-625 filler-wire could preferentially absorb most of the laser energy, while the residual energy was sufficient to melt part of the CM247LC. The energy was insufficient to mix it with the IN-625 filler wire, thus leaving an uncertain state. Following hot isostatic pressing, particles which were sub-micron or nanometres in size existed along the healing trace. As compared with the surrounding matrix, these particles were rich in hafnium and tantalum, suggesting that the precipitates were (Hf,Ta)-rich carbides.

The CM247LC is especially sensitive to weld-cracking during powder-bed fusion which is caused by an unsuitable laser-spot diameter resulting from focus-shifting. The effect of a 4mm focus-shift upon melt-pool geometry, crack-density, microstructure and texture was investigated[69]. So-called hot calibration was used to overcome a thermally-induced focus-shift of 4mm by setting the focal plane 4mm below the build plane. The procedure permitted fully dense and crack-free CM247LC samples to be prepared. Due to deeper melt-pools, produced by a higher intensity and a reduced laser-beam diameter in the focal plane, cracks in the under-layers were re-melted and crack-healing occurred. Samples which were produced using the hot calibration system had a stronger texture parallel to the build direction. With regard to crack-density, the number of cracks per unit area was reduced by more than 98 times by the hot calibrated laser system and finally resulted in crack-free samples. The crack-healing occurred only in layers below. Using the thermal steady-state laser system, the melt pool depth increased from 59 to 81μm, resulting in a roughly 37% deeper melt-pools, as compared to those in cold-calibrated samples. This was due to the higher-intensity input, due to a smaller laser-beam diameter of 90μm in the focus plane, as compared to a 116μm beam diameter in the defocused case of cold-calibration. Due to the deeper melt-pools, cracks were re-melted in the layers below, and crack-healing could occur; resulting in a 99.96% relative density and crack-free CM247LC samples. The laser-beam diameter was identified as being an essential parameter for crack-healing. A hot calibrated system, with a smaller beam-diameter in the focal plane and a higher intensity input, produced a stronger texture parallel to the build direction as compared to that of a cold-calibrated system with a weaker texture.

Nickel-based superalloys such as CM247LC are prone to cracking when prepared using laser powder-bed fusion. The cracking in large-scale applications is due to thermal energy accumulation and low thermal conductivity. Post-processing methods such as hot isostatic pressing can mitigate cracking, but its effectiveness may be reduced in material which is produced under high volumetric energy-density conditions. A study was made[70] of the effect of impact of hot isostatic pressing on CM247LC which was fabricated at volumetric energy-densities of 43.65 to $159.72 J/mm^3$. The crack-healing effect of hot isostatic pressing under low and high volumetric energy-density conditions was about 90 and 47%, respectively. The cracking mechanisms operating during manufacture were cracking (hot-tearing) in the last phases of solidification when solute-rich liquid phase was trapped between solidified interfaces, and liquid-to-solid transformation-induced shrinkage which caused residual liquid to be pulled apart. The resultant cracks tended to nucleate within grain cores under high energy-density conditions. Grain-boundary (liquation) cracking occurred within the heat-affected zone of a molten track. Microsegregation during additive manufacture reduced the solidus temperature of grain boundaries and interdendritic regions which became liable to tearing. Linear cracks formed under low energy-density conditions, were generally aligned with grain boundaries and almost parallel to the building direction. The volumetric energy-density marked the main difference between the cracking mechanisms. The solidification and liquation cracking occurred mainly at high and low volumetric energy-densities, respectively. Crack-density measurements were used to assess the effect of hot isostatic pressure pressing upon the 2 types of cracking. The average crack-density of as-prepared samples was 0.08%; equating to an 89.3% reduction in crack-density. The average crack-density of samples which were fabricated by using a high volumetric energy-density decreased from 1.99 to 1.05% after hot isostatic pressure, equating to a crack-reduction of 47.2%. This indicated that the pressure treatment of samples that contained mainly liquation cracks offered more effective crack-healing than it did in the case of solidification cracks. The main mechanism of healing of the former cracks involved the elimination of grain-boundary segregation. The application of high temperatures and pressures induced elemental diffusion and reduced the interfacial potential energy on both sides of cracks, led to the transfer of solutes in the grain boundary. Solidification cracks could not be closed by solute-transfer and elemental diffusion during pressing, because they formed mainly across the grain core during fabrication. These cracks did not have much solute at the crack boundaries, and this hindered the solute-transfer which depended upon the presence of a solute-rich liquid phase. The irregular morphology of solidification cracks hindered solute from bridging crack surfaces and impeded the process of crack closure. The 47.2% reduction in crack-density of samples fabricated by using a high volumetric energy-density was attributed to incomplete closure of the solidification cracks, due to insufficient solute-transfer. The main reason for the density-reduction was partial reduction of the size of solidification cracks, rather than complete closure. The observed crack-healing was attributed to the closure of liquation-cracks, some of which were present in specimens with a mainly solidification-cracked morphology. This suggested that even a small fraction of liquation-cracks could contribute to overall crack-healing during hot isostatic pressing. Specimens which were fabricated by using a low volumetric energy-density exhibited incomplete crack-healing following pressing, due to remaining unclosed solidification cracks. Specimens which were prepared by using a low energy-density had an average relative density of 99.92% following pressing. The use of a lower energy-density guaranteed sufficient crack-healing during hot isostatic pressing. This also avoided the occurrence of thermal defects which were caused by the reduced accumulation of thermal energy during manufacture.

Palladium

Compatibility-induced stresses can develop in polycrystalline materials regardless of grain-size, but are particularly important in nanocrystalline materials because such stresses are less easily alleviated by dislocation-based plasticity. As self-healing mechanisms help to eliminate small incipient cracks, materials having a higher self-healing potential can offer a greater resistance to crack-initiation during fatigue. Some nanocrystalline metals are particularly fatigue-resistant and cracks are initiated only after local grain-coarsening. This suggested that the study of fatigue in nanocrystalline systems was a suitable test-bed for exploring crack-healing mechanisms. Molecular dynamics simulations were therefore made[71] of the crack-healing which was produced by the external mechanical loading of 3-dimensional models of nanocrystalline palladium. The healing behaviour was due to the interaction of the crack with internal stresses in the surrounding microstructure. The results illustrated the importance of microstructurally-induced internal stresses when predicting the behaviour of nano-cracks under mechanical loading. Compressive loading usually promotes crack-closure, thus creating favourable conditions for bonding and healing. It is somewhat counter-intuitive that monotonic shear and tensile loadings can also induce crack-healing if those loads generate non-uniform internal stresses that produce localized compression near to the crack-site. Previous molecular dynamics simulations of 2-dimensional non-periodic microstructures had shown that stress-driven grain-boundary migration can lead to localized compression at crack-tips, possibly leading to crack-healing. This idea was extended to more realistic 3-dimensional periodic models of nanocrystalline palladium by using large-scale molecular dynamics simulations. Crack-closure and healing occurred under simple shear loading in the nanocrystalline material, although no nett compressive stress was applied. A further simulation was performed, involving a monocrystal containing an identical crack which was oriented so as to match the grain in the nanocrystalline model. All of the other simulation parameters were left unchanged. Under a shear stress of about 1.5GPa and a shear strain of 4.4%, Shockley partial dislocations nucleated at the crack and propagated in its plane. No crack-closure occurs. Differing crack-behaviours which were observed in the nanocrystalline and monocrystalline models were attributed to the non-uniform internal stresses which were generated by microstructural heterogeneity in the nanocrystalline case.

In order to investigate how the stresses arose, a further simulation was made of the same nanocrystalline microstructure, but with no crack. Atomic-level virial stresses were calculated for a shear strain of 4.0%. This was just below the point at which crack-closure began in the original simulation. The atomic-level stresses were calculated by using the atomic volume of a stress-free face-centred cubic palladium crystal. This analysis included only stresses from grain interiors, and excluded those near to grain boundaries and surfaces.

The results revealed that complete crack-closure and healing could occur in realistic 3-dimensional periodic microstructures, even in the absence of an externally-applied compressive stress. It was confirmed that the mechanism of mechanically-induced crack-healing which had previously been proposed for 2-dimensional non-periodic models was not an artefact of the simplified geometry. That is, it is not driven by surface image-stresses, but by internal stresses which resulted from evolving microstructures which arose from compatibility-strains and inelastic grain-boundary responses. It is therefore necessary to account for microstructurally-induced internal stresses when predicting crack behaviours under mechanical loadings. A wide range of crack responses can occur, depending upon the grain shape, orientation and elastic properties near to the crack.

The role played by local tensile stresses was particularly important. Martensitic transformations and differential distortions can also generate internal stresses and thus influence crack-behaviour. It remained unclear whether the present small-scale crack-healing events could be exploited in order to improve material performance. That is, could mechanically-imposed crack-healing be achieved in polycrystalline material by controlling grain-size, texture and grain-boundary orientation-distributions?

Silver

Photonic curing was applied to aqueous conductive inks which consisted of micron-sized silver flakes. This was found to increase the conductivity beyond that possible by using thermal curing alone. It was demonstrated[72] that the photonic curing partially pre-melted silver flakes in the ink. This then led to necking between adjacent particles and to the healing of cracks. The resistivity of screen-printed fine lines changed from an average of $1100\mu\Omega$cm for thermal curing alone to an average of $100\mu\Omega$cm with the help of photonic curing. The most conductive patterns had a resistivity of $60\mu\Omega$cm after photonic curing, corresponding to a conductivity of 1.7×10^7S/m. The yield of conductive fine lines increased from 52% for thermal curing to 80% for photonic curing alone and to 100% for optimum photonic-plus-thermal curing. This was all explained by a crack-healing process due to the photonic curing, during which the observed interparticle necking was able to bridge cracks in the samples. Given that very fine cracks were a common failure mode during the manufacture of printed lines, the curing effect permitted many non-conductive structures to be repaired in-line. In further analysis, samples were grouped into uncured and thermal alone and photonic plus thermal plus photonic. The mean particle-size was greater in photonic-treated samples than in untreated samples, increasing from 1.8 to 2.10μm, respectively. This was consistent with the assumption that photonic curing caused localized pre-melting at the flake edges. The size distribution of flakes was skewed towards larger flakes in photonic-treated samples and towards smaller flakes in untreated samples. This was consistent with a process which caused smaller particle to fuse into larger particles with about double the length. The increase in mean particle-size strongly supported the crack-healing hypothesis.

Tin

A 800μm-diameter SAC305 (Sn-3Ag-0.5wt%Cu) solder bump was soldered to a substrate, and resistance changes were monitored[73] during current aging. The joint-resistance before aging was 7.7 and 11.8mΩ at room temperature and 160C, respectively. The resistance of the solder joint increased instantly by about 1mΩ when subjected to a 2.2A aging current at 160C. The increase was gradual for some hours, and was sharper upon approaching final failure. Four stages could be identified, and correlated with cross-section observations. The stages were intermetallic-compound growth, crack formation and propagation, intermittent crack healing-forming and final failure. A periodic fall and rise was noted and was attributed to the intermittent crack healing-forming stage. The healing events were faster than the sampling time. It was proposed that, as the current was concentrated when by-passing interfacial cracks, local melting occurred and partially filled cracks before re-solidifying. Although crack formation and healing were possible in stage-1, it seemed to be too small to affecting the resistance signal. Some cracks were caused by electromigration and, due to the directional nature of electromigration, cracks formed preferentially at the cathode. An apparent instantaneous drop in resistance was attributed either to rapid atomic diffusion filling a crack or to local melting and crack-filling with molten material caused by increased Joule-heating of interfacial asperities next to the crack. When the joint-resistance increased to a critical value, due to crack propagation, the Joule-heat increased locally due to the current being concentrated. When the Joule-heat was so high that the local temperature reached the melting point of the alloy, the latter melted locally at the interface and filled nearby cracks. Less of the current passed through the molten part, due to the higher resistivity of the liquid metal as compared to that of the solid. The local temperature thus fell suddenly and the molten metal solidified.

Titanium

The use of electropulsing to heal fatigue-induced cracks, and to improve the fatigue resistance, of single edge-notched pure titanium was studied[74]. Fatigue cycling was used to introduce a pre-crack into the notch root. A pulse with a current density of $88A/mm^2$ and a duration of 0.4s was then applied to the fatigued specimen. This showed that the treatment completely healed the fatigue-induced crack and extended the fatigue life of the specimen by more than 1000000 cycles from the original life of about 55000 cycles. Electro-thermal simulation and microstructural analysis showed that the great increase in fatigue life was due mainly to the generation of high compressive residual stresses which resulted from inhomogeneous thermal expansion and from phase transformation-induced volume changes in the notch region. Microstructural changes during the treatment lowered the textural strength and increased the ductility of the notch root, thus further increasing the fatigue life. The compressive residual stresses and the increased ductility at the notch-root inhibited crack-initiation and propagation. The effectiveness of the electropulse treatment in healing the fatigue-induced cracks was then evaluated by continuing the fatigue test up to fracture or until one million fatigue cycles were completed. The average fatigue life of single edge notched specimens was 55573 cycles. The specimen was subjected to 33344 fatigue cycles before electropulse treatment. This was equivalent to about 60% of the average fatigue life of the material. A 300µm-long crack nucleated at the notch-root and propagated perpendicularly to the loading direction. In the absence of electropulsing, continued fatigue-testing after the interruption at 33344 cycles was expected to cause failure after 22229 more cycles. If electropulsing simply eliminated the crack, failure was expected to occur at about 55573 cycles. Resumed fatigue tests instead demonstrated that an intermediate electropulse treatment not only healed the crack but also markedly extended the fatigue life. When such tests were stopped after one million cycles, no fatigue failure was observed and no cracks were found at the notch root. The mechanism of electropulse crack-healing was proposed to be solid-state joining, with enhanced diffusion. When the electric current is applied to a pre-cracked specimen, it detours around the crack-tip and concentrate at that location. This markedly increases the temperature at the tip, generates Joule heating, produces higher compressive stresses and leads to crack-healing via boundary-migration, cavity-entrapment and partial substructure alignment. Another possible crack-healing mechanism was however local micro-welding due to melting, rather than solid-state joining. Such micro-welding would result in a solidified microstructure around the crack-tip and re-fill the fatigue crack. The effect of surface oxidation upon the fatigue behaviour was not considered. Because intermediate electropulse treatment healed fatigue-induced cracks and extended the fatigue life, the treatment was also expected to be a useful pre-strengthening technique. In further tests, the same electric pulse which was used for crack-healing was applied to an unfatigued specimen in order to pre-strengthen a notched specimen.

Laminated titanium/aluminium sheets were prepared[75] via the hot-pressing, under 20MPa, of pure titanium and aluminium foils at 630C for various times. The thickness of $TiAl_3$ layers increased and the elongation of Ti/Al laminated sheet decreased with hot-pressing time. Sheets which were hot-pressed for 0.25h exhibited the greatest elongation: 135% at 600C. Crack initiation and propagation in the hot-pressed sheets were studied by tensile-testing to various strains. When the temperature was between ambient and 400C, the cracks in the $TiAl_3$ layers were mutually parallel, and were perpendicular to the tensile direction. The size and number of cracks increased with increasing strain. At 600C, parallel cracks appeared after 70% deformation. After 100% strain, the $TiAl_3$ layer-thickness markedly increased, with much disordered and discontinuous cracking. An improved sheet formability was observed because the cracks were filled, and healed, by newly-formed $TiAl_3$ during deformation at 600C. The $TiAl_3$ layer was maintained for longer before rupturing, thus greatly improving the coordinated deformation of various layers. The fracture surfaces at room-temperature, 200C and 400C exhibited delamination and necking. Larger and deeper dimples appeared in the titanium and aluminium layers at 600C, reflecting an excellent ductility plus coordinated deformation. The plasticity of the 0.25h hot-pressed sheet increased with temperature. Vertical cracks appeared in the $TiAl_3$ layers at 10, 10 and 40%, at room-temperature, 200 and 400C, respectively. No cracks propagated into the titanium or aluminium layers. At 600C, vertical cracks appeared after 70% deformation. When 0.25h hot-pressed sheet was subjected to hot-gas bulging tests, it had a bulge-height of 27mm with a height/radius ratio of 0.9. The thickness of the titanium and aluminium layers increased from top to bottom. In the top region, the TiAl$_3$ layers broke into spherical fragments, and the thickness increased. During tensile deformation at 600C, the TiAl$_3$ layer thickness was much greater than its predicted value, indicating that the tensile stress promoted atom-diffusion.

The effect of healing upon the mechanical properties of titanium alloy was investigated[76] by applying pulsed-current and laser treatments to deep cracks. Microhardness data showed that, following electropulse treatment, the hardness difference between the substrate and healed areas was significant, with a greater hardness being observed in the healed areas. Following laser-remelting, the hardness of the healed areas which were generated by electropulsing was reduced. Following electropulse treatment, the tensile properties decreased and the fatigue properties improved. In the case of specimens which were subjected to combined pulsed-current and laser treatment, the tensile and fatigue properties were both improved. The Joule-heating effect of the pulsed currents, and the re-melting effect of lasers were the main causes of crack healing.

Combined pulsed-current and laser treatments have been applied[77] to deep cracks in titanium alloy. During the pulsed-current treatment, high temperatures and local compressive stresses were generated at crack tips. Some healed regions appeared following pulse-current treatment, and electron microprobe results showed that the distribution of elements in the healed regions was uniform while electron microprobe analysis showed that the microstructure comprised mainly fine acicular martensite. The healing of deep cracks was evident following combined treatments. Laser re-melting resulted in the formation of a re-melted zone and a heat-affected-zone structure. At points near to the re-melted zone, the healed areas which were produced by pulsed currents exhibited a largely columnar structure.

Ti-Al

Laser shock-peening can repair surface damage and thereby improve the safety and reliability of titanium alloys. In the case of Ti-6wt%Al, crack evolution for various orientations under laser-shock conditions was studied[78] by using molecular dynamics simulations. The plastic deformation mechanism of the titanium alloy exhibited an obvious orientational correlation. This was dominated by a partial pyramidal dislocation and short stacking fault for the [00•1] orientation model, and by crystalline reorientation accompanied by a so-called sandwich structure for the [10•0] orientation model. The structure was superimposed by atomic shuffling and reverse glide on the basal plane, consistent with the geometric deformation in the impact direction and vertical direction. In the [10•1] orientation model, phase transformation was induced by multiple stacking faults. The healing times of micro-cracks having differing orientations were closely related to the compression stress and dislocation activation. Healing was easiest for the [10•1] orientation crack, due to the nucleation and accumulation of multiple dislocations at the crack surface. The tensile strength and strain of polycrystalline models, following crack healing, were greatly improved.

Ti-Al-Nb

The Ti-43.5Al-4Nb-1Mo-0.1B alloy tends to suffer from surface-cracking during processing and use. A new electroshock treatment was used[79] for surface-crack repair and strengthening. Nearly *in situ* experiments and theoretical analysis revealed the effect of various current densities and electroshock techniques upon the healing of surface cracks. Scanning electron microscopy showed that the principal factors which affected the healing were the time of energization, the current density and the crack angle. The crack-tip and locally affected regions were more likely to be healed quickly under continuous electroshock treatment with a current density of $75A/mm^2$. The surface-crack width decreased from 0.47μm to 0μm. Electron back-scatter diffraction data showed that the dislocation density increased around the surface-crack healing area; the dislocation-density at the crack-tip, and away from the healed region, decreased and was uniformly distributed following electroshock treatment. This could further suppress the generation and propagation of a crack, leading to a 20.1% increase in tensile strength and to a 40.3% increase in fracture strain. By assuming a coupling of the thermal and non-thermal effects of the electroshock treatment, a 2-stage mechanism involving thermal compressive stresses and crack-bonding interfaces could describe the evolution surface-crack healing during electroshock treatment. Large cracks, those having widths which exceeded 0.5μm, were difficult to repair when the applied current-density was low. When using a moderate current-density, the crack-healing capability of the electroshock treatment markedly improved. When a high current-density was applied, further crack-width reduction occurred. With a constant energy input, the treatment had a notable surface-crack repair effect, due mainly to crack-edge passivation and localized healing. Slender cracks exhibited greater healing when compared to that of almost circular micro-cracks having similar aspect-ratios. Within the optimum range of current-density, the electroshock treatment produced effective crack-healing, but an excessive current-density could widen cracks. A two-stage atomic diffusion-based healing mechanism operated during the treatment. The alloy initially featured V-shaped surface cracks which formed during manufacture, and this region often exhibited a high dislocation-density, especially at the crack boundaries. This increased atomic diffusion. Cracks with sharp V-shaped geometries, and bulging defects, were particularly prone to heal when the electric current during electroshock treatment flowed perpendicularly to them. During the first stage of treatment, current diversion around the crack-tip and defected area occurred. This diversion then led to a localized Joule heating that increased the temperature at those locations while the surrounding matrix remained relatively cool. The resultant temperature gradients created compressive stresses at the crack-tips. The concentration of strain and dislocations at the crack-tip created an efficient pathway for atom-migration. The various fields which were generated by the process accelerated atom and vacancy movement towards bonding interfaces and grain boundaries. Because thermal compressive stresses promoted interface adhesion, the electroshock treatment further drove atom-diffusion from high to low current-density zones. During the second stage, atoms from high-temperature regions continued to diffuse and finally established an homogeneous elemental distribution within the healed area. This process aided moreover the annihilation of opposing dislocations near to, and away from, the crack; thus decreasing the dislocation density. When the dislocations were impeded by boundaries, their accumulation led to localized increases in the dislocation density. Thermal compressive stresses mainly drove crack faces together in the early stage. As the healing progressed, atoms and vacancies diffused across the bonding interface and progressively repaired the crack. The crack-tip was passivated first, with healing then gradually extending to the remainder of the damaged region. The progression from crack-tip passivation to complete healing improved the mechanical properties of the material by reducing the dislocation density and reinforcing the structure near to the original crack-site.

The 3-point bend strength, Young's modulus and vibrational damping of a plasma-sprayed molybdenum-coated Ti-25Al-10Nb-3V-1Mo alloy were measured[80], revealing that the bend strength was markedly affected by the molybdenum coating. The decrease in strength was attributed to the presence of cracks which formed in the molybdenum coating during plasma-spraying. Piezoelectric ultrasonic results indicated that the modulus and vibrational damping of coated samples were much higher than those for non-coated substrates. Thermal cycling of the molybdenum-coated material between 600C and room temperature revealed that there was an increase in the modulus, and a decrease in the mechanical damping. The modulus of the coated samples was notably higher than that of the uncoated ones. This was consistent with the higher modulus of molybdenum. This increase was accompanied by a reduction in the mechanical damping, especially following one thermal cycle, suggesting that a healing mechanism was operating in the material. In the molybdenum-coated intermetallic samples, an initial thermal cycle led to a marked increase in the modulus and to a corresponding decrease in damping, confirming crack-healing. Scanning electron microscopy also confirmed that cracks which had formed during plasma-spraying had healed during heat exposure. This healing was suggested to be aided by the presence of traces of aluminium in the coating, and was associated with considerable grain growth. The microstructural changes directly contributed to the enhanced modulus and reduced damping. In spite of an absence of post-cycling bend-strength tests, it was inferred that the strength of thermally-cycled coated samples would be comparable to that of uncoated ones; given the mitigating effect of crack-healing upon strength degradation. On the other hand, failure of all of the coated samples originated from the coating on the tensile side during bending. This initiated interfacial flaws between the coating and the substrate, and finally led to cleavage fracture of the intermetallic substrate. The higher modulus of molybdenum (280 to 330GPa), relative to that of the substrate (110 to 125GPa), accounted for the increased stiffness of the coated samples. The higher damping which was observed in coated samples was initially attributed to micro-cracks in the coating. Uncoated samples exhibited only a marginal increase in modulus during thermal cycling.

Ti-Al-Sn

Hot-compression was used to investigate the effect of temperature on crack-healing in Ti-6Al-2Sn-2Zr-3Mo-1Cr-2Nb-Si alloy (TC21)[81]. The temperature greatly affected the crack-healing process, and the extent of crack-healing increased with increasing temperature. This improved the degree of void-closure because the proportion of β-phase, and the atomic diffusion coefficients, increased with increasing healing temperature. The velocity of grain boundaries also increased with increasing temperature, leading to a greater migration distance. This contributed to the complete crack-healing of this alloy. Optical microscopy near to crack interfaces following hot-compression healing at various temperatures revealed voids, and a microstructure which comprised white α-phase and dark β-phase. At 850C the crack-healing was minimal, with voids which had a length of 23.1 to 73.4µm and a width of 2.4 to 6.6µm; indicating insufficient healing (figure 14). Increasing the temperature to 900C led to a transformation of the voids into smoother and more rectangular shapes which were 4.4 to 16.5µm long and 1.9 to 4.3µm wide; suggesting improved healing. At 950C, many small round or oval voids formed which had a length of 3.0 to 9.1µm and a width of 1.7 to 3.8µm. The maximum lengths and widths of voids fell sharply between 850 and 900C, followed by a lesser decline between 900 and 950C; suggesting that higher temperatures enhanced crack-healing in this alloy. At 850C, the crack interface remained planar, with small (0.96 to 2.28µm) nucleated grains which did not span the crack. At 900C, grain-growth continued and a few grains began bridge the crack. At 950C, further grain-growth led to an interface-bending which was attributed to grain-boundary migration. As the temperature increased, the internal

Materials Research Forum LLC

https://doi.org/10.21741/9781644903773

misorientation grew and attained some 10° at 950C, suggesting the occurrence of continuous dynamic recrystallization via sub-boundary rotation and the formation of new high-angle boundaries. At 850C, many high dislocation-density and some low dislocation-density zones were observed. Sub-boundaries formed via dislocation rearrangement. Dislocation-free grains were present at the interface, reflecting discontinuous dynamic recrystallization. At 900C, the density patterns remained similar. At 950C, the high-density regions shrank and the grain-size increased.

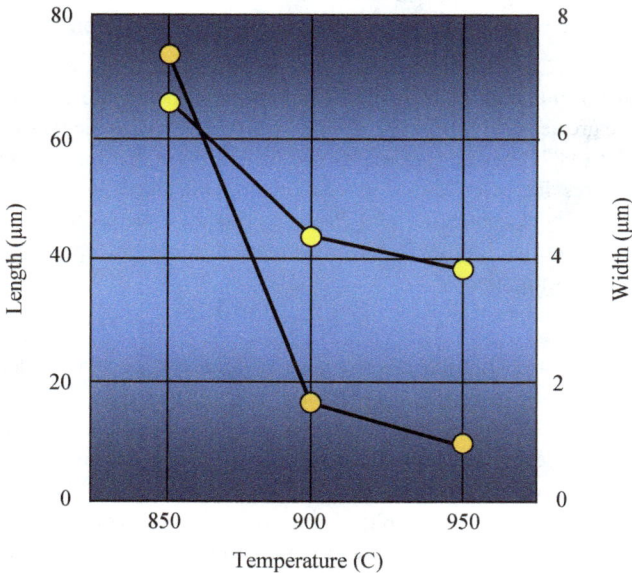

Figure 14. Maximum lengths and widths of voids in TC21 healed at various temperatures. Yellow: width, orange: length

An increased healing temperature led to a reduction in the number of low-angle grain boundaries and to an increase in the number of high-angle grain boundaries, especially from 900 to 950C. Recrystallized and sub-grains increased while deformed grains decreased; particularly at 950°C. During healing, initially-long strip-like voids at 850C, with irregular edges, transitioned smoothed into oblong voids at 900C and later into small round voids at 950C, which finally disappeared. These changes were driven by surface diffusion, with atoms migrating from regions of higher curvature to lower, due to differences in chemical potential.

Continuous dynamic recrystallization occurred gradually via sub-grain rotation and boundary formation when the misorientation was greater than 15°. Discontinuous dynamic recrystallization involved nucleation at bulged grain boundaries, followed by rapid growth. Its nucleation was energetically favourable at lower strain thresholds, and thus predominated at low healing pressures. Its nuclei, which originated mainly at crack interface, drove crack-closure via accelerated boundary-migration. This could be 3 orders-of-magnitude faster than continuous dynamic recrystallization.

Crack-healing in TC21 alloy thus involved 2 main stages: void-closure and boundary-migration. The former was aided by plastic deformation and atomic diffusion and both of the latter were greater at higher temperatures. These were near to the $\alpha \rightarrow \beta$ transformation range of 873 to 982C, where β-phase formation increased ductility. At the same time, high temperatures increased diffusion coefficients and aided atomic mobility around voids. In general, both continuous and discontinuous dynamic recrystallization played a role in the crack-healing, but the latter predominated due to its rapid kinetics and more effective boundary-migration.

Ti-Al-V

Pulsed-current assisted forming can be used to improve the plastic deformability of titanium alloys. Shear-testing of Ti-6Al-4V was performed[82] by using hat-shaped specimens under pulsed-current and constant-temperature conditions. The deformation in the shear-zone under electroplastic shearing conditions was greater than that under isothermal shearing conditions. The shear load was decreased by the pulsed current. A clear straight shear-band was observed in electroplastically-sheared specimens. Intracrystalline deformation of grains was greatly facilitated by the pulsed current, with greater deformation of the grains occurring along the shear direction. Micro-cracks were present in the shear-zone under isothermal shearing conditions, but none were found in the shear-zone under electroplastic shearing conditions. Clear crack-healing occurred at the crack-tip of the shear-zone under electroplastic shearing conditions.

The micro-cracks formed along the grain boundaries when intergranular deformation exceeded a threshold level. The application of a pulsed current led to the immediate suppression and healing of micro-cracks. The resistance and temperature in the cracks increased rapidly due to the pulsed current, leading to crack closure imposed by hydrostatic pressures acting perpendicularly to the shear direction, followed by welding due to Joule heating. Previous studies had identified Joule heating and hydrostatic pressure on crack interfaces as being key mechanisms in crack healing. Fracture crack-

Materials Research Forum LLC

https://doi.org/10.21741/9781644903773

tips during electroplastic shearing were blunter than those during isothermal shearing. Observations revealed clear evidence of crack-welding, and analysis confirmed the presence of oxidation at high temperatures during the process.

The effect of electropulse-assisted ultrasonic surface rolling (table 9) upon the surface mechanical properties of Ti-6Al-4V was studied[83]. The introduction of optimum electropulsing facilitated surface-crack healing, and improved the surface microhardness and wear-resistance. The residual compressive stress was further increased by the enhanced electropulse-assisted process. The rapid improvement in surface mechanical properties was attributed to ultra-refined grains and to an increased plastic deformation due to the coupling of ultrasonic rolling and electropulsing. The healing of surface cracks became more marked with increasing frequency and specimens which were treated at 350Hz exhibited an excellent surface morphology, with few cracks and defects being visible.

Non-electropulse treatments could reduce surface roughness, but the processed surface could deteriorate somewhat due to shear deformation and local fatigue damage under excessive processing, which favoured a decrease in deformation resistance. In the electropulsed case the treatment could encourage micro-crack healing and reduce the number of defects, thus further improving the surface quality via an increase in dislocation mobility, atomic migration and plastic deformation within the surface layer. The microhardness was typically 300HV before processing. Following treatment, there was a marked improvement in surface hardness. In the case of the non-electropulse treatment, the hardness at the top surface increased from an initial value of 385HV and the impact depth was some 300mm.

When electropulsing was added, the hardness at the top surface decreased, following an initial increase, as the frequency was increased. The impact depth of the strengthened layer exhibited the same variations. At 250Hz, the hardness of the top surface of electropulsed specimens attained 423HV and the cross-sectional microhardness was higher than that found in the absence of electropulsing. The great improvement in surface hardness, and a decrease in the friction coefficient, which were observed in the presence of electropulsing were deemed to be due to crack-healing. The high strain-rate provided by ultrasonic rolling, the increased dislocation-mobility and the atom diffusion which was induced by electropulsing were concluded to be the primary reasons for the observed behaviour.

Table 9. Electropulsed ultrasonic surface rolling parameters for Ti-Al-V

Frequency(Hz)	RMS Current Density (A/mm^2)	Surface Temperature(C)
200	0.72	108
250	0.86	131
300	0.95	153
350	1.07	171

TiZrNbV

The dynamic compressive behaviour of TiZrNbV body-centred cubic high-entropy alloys was studied, and a notable flow-stress recovery was observed[84] in dynamic compressive stress-strain curves which were determined by using split Hopkinson pressure bar tests. A crack-healing phenomenon was found to occur in the TiZrNbV during dynamic compression, and thus led to the recovery of the flow stress. It was suggested that explosive-welding dynamic compression led to the crack-healing behaviour. Micro-crack initiation occurred when the strain exceeded 8%, and the cracks formed mainly along grain boundaries.

As the strain increased, primary cracks widened and some secondary micro-cracks began to propagate into the grain interiors. At strains of 27% and 37%, crack-healing appeared between two prominent primary cracks. With further straining, partial healing at the crack-tips of some main cracks appeared. At a strain of 6%, grains exhibited an elongation that was aligned with the loading direction. Fractured fine grains were present in material which was strained to 6% and 8%. When the strain increased to 12%, there was a notable change in the grain size.

During dynamic compression, Ti$_{18.5}$Zr$_{48}$Nb$_{18.5}$V$_{15}$ and Ti$_{23.9}$Zr$_{17.2}$Nb$_{23.9}$V$_{35}$ exhibited only a slight lattice distortion. An increased strain-rate and a greater deformation which were observed during dynamic compression were attributed to the discontinuous nature of the applied pressure. A higher pressure, corresponding to a higher external load, led to an increased kinetic energy and impulse. So while TiZrNbV high-entropy alloys which were strained at a rate of 7400/s absorbed enough energy to enable explosive welding and crack-healing, alloys which were subjected to strain-rates of 5000 and 5900/s did not absorb enough energy. The critical deformation energy was estimated on the basis of lattice distortion and was related to mechanical and thermal properties.

Tungsten

Density measurements were made[85] of 0.762mm $W-1wt\%ThO_2$ wires, following annealing at between 1000 and 2000C for up to 2h. The observation of a low as-worked wire-density was attributed to large cracks which were associated with the thoria particles. Time-temperature annealing conditions, which were very much below which were required for recrystallization of the wire, resulted in partial healing of the cracks. The crack-healing process could be followed by monitoring longitudinal fracture surfaces and the wire density. The apparent activation energy for crack-healing was 57.8kcal/mol. The results suggested that crack-closure was a result of shear and grain-boundary diffusion-transport. It was noted that the presence of cracks could affect the morphology of recrystallized grains. The crack-healing behaviour and microstructural evolution of pure tungsten, produced by laser powder-bed fusion, were compared[86] before and after hot isostatic pressing. The thermal conductivity was 133W/mK at room temperature, following the treatment. This was 16% higher than that (115W/mK) of as-prepared samples. The treatment had little effect upon the density, but it led to a grain-size of more than 300μm and a decrease in dislocation density and crack-healing, which led to an appreciable improvement in the thermal conductivity.

Zirconium

Molecular dynamics simulations were used[87] to investigate interactions between collision-cascades and nano-cracks. When a thermal spike overlapped with a nano-crack, the collision-cascade could induce healing of the crack. A higher primary knock-on atom energy led to a higher degree of crack-healing for a given separation distance between the primary knock-on atom and the nano-crack. The primary knock-on atom velocity direction changed the fraction of atoms which entered the crack (table 10), by affecting the shape and distribution of the thermal spike. The degree of crack-healing decreased as the distance between the nano-crack and the primary knock-on atom decreased. Collision-cascades effected the transformation of a nano-crack in the basal plane into a prismatic vacancy loop due to the interaction between the thermal spike and nano-crack. For example, the response of a collision-cascade, triggered by a 10keV primary knock-on atom was modelled by using various time-steps. The primary knock-on atom was positioned 32Å from the crack surface and was directed perpendicularly to the [00•1] orientation. After 0.02ps, the primary knock-on atom reached the nano-crack surface and defect atoms were dispersed along the [00•1] direction. After 1.06ps, the system attained a thermal-spike state, with most of the high-temperature region overlapping the crack. During cooling, diffusion led to the recombination of interstitials and vacancies. In the final stable state, numerous point defects and clusters were present in the matrix and the nano-crack was almost entirely healed; indicating much crack-recovery due to cascade effects. Healing was confined to regions within which the thermal spike overlapped the crack. In another example, the primary knock-on atom was located 105Å from the crack and was also directed perpendicularly.

Table 10. Fraction of atoms entering a crack, following a collision-cascade, versus the primary knock-on energy; the distance between the crack surface and the primary knock-on atom being 105Å.

Primary Knock-On Atom Energy(keV)	Fraction of Atoms Entering Crack(%)
10	10.5
15	19.9
20	45.8
30	63.2

Here, the thermal spike did not overlap with the nano-crack. After 22.35ps, the nano-crack remained clearly visible, with residual point-defects; indicating no significant healing. This revealed the important role played by thermal-spike overlap in promoting crack-healing. In order to quantify the degree of healing, calculations were made of the ratio of atoms entering the crack region, to the initial number of atoms removed during crack-formation. At the peak of the thermal spike, the degree of healing attained some 70% and approached 100% in the final stable state. The simulations took account of the effect of various parameters, such as the primary knock-on atom energy (10keV), direction ([00•$\bar{1}$]), crack-thickness and separation-distance (32, 79 or 105Å). In every case, the healing efficiency decreased as the separation-distance increased from 32 to 105Å, regardless of the crack-thickness and primary knock-on atom direction. Thinner cracks exhibited better healing for given separations and directions. For a given separation, atom in-flux varied markedly with primary knock-on atom direction. Two main mechanisms were considered. In one proposal, the thermal spike was supposed to emit compressive-stress waves which affected the surrounding material. One simulation indicated the existence of increased stresses in the thermal-spike core. Other studies had suggested however that stress alone was insufficient for crack closure. Another proposal was that collision-cascades could increase local temperatures to far above the melting-point of zirconium. One simulation indicated temperatures in the spike-core could attain 5000K, so that the material could liquefy for 1 to 2ps and the liquid could enter the crack via diffusion or pressure-driven flow. The simulations showed that, when full overlap occurred, healing attained about 70% after 1.06ps and about 90% after 2ps, thus supporting the role of thermal-spike dynamics in nano-crack healing.

Molecular dynamics simulations were again used[88] to make a similar investigation of the interaction between micro-cracks and overlapping collision-cascades. Crack-healing occurred when the core of the thermal spike covered the crack, due to the recrystallization of atoms of high kinetic energy following the first collision-cascade. The formation of vacancy dislocation loops, in crack-healed material with overlapping collision-cascades, accounted for the combined effects of overlapping collision-induced vacancy collapse and recrystallization-induced vacancy migration.

Uniaxial tensile deformation of crack-free and cracked material, before and after irradiation, revealed strength-enhancement following crack-healing, an inhibition of pyramidal slip and promotion of the formation of nano-twins. The movement of nano-twin planes controlled dislocation nucleation and movement. Irradiation-induced crack-healing markedly altered the crack morphology, and influenced the evolution and distribution of defects in cracked zirconium. Cross-sectional simulated views of microstructural evolution 0, 0.2, 0.9, 5.0, 20.0 and 30.0ps after the first collision cascade showed that knock-on atoms reached the crack region within 0.2ps and initiated a thermal spike which overlapped the crack and resulted in localized melting and atomic mixing in the thermal-spike core. As the collision-cascade ended, atoms within the cascade-core cooled and recrystallized, contributing to crack-healing and to the formation of stacking-faults around the crack.

In order to assess quantitatively the extent of crack-healing, account was taken of the numbers of atoms entering the crack and of the corresponding healing-rate. The latter increased by 86.9% within 2.9ps and then continued to rise gradually until it reached a steady-state peak of 94%. Microstructural observations indicated that crystallization within the crack region began at 5ps, with maximum crystallinity occurring after about 25ps. During recombination, the fraction of hexagonal close-packed atoms increased at the expense of amorphous phase and stabilized after 25ps. Increased crystallization improved phase-stability and promoted the self-healing ability of cracked zirconium. The numbers of atoms which entered the crack, and the healing-rate following successive cascades were also monitored. Following the first cascade, 394 zirconium atoms penetrated the crack region and increased the healing-rate to 94%. Following the second cascade, the crack was fully-healed and attained a healing-rate of unity. Minor reductions in healing-rate could occur following subsequent cascades, but it remained as high as 94%. The slight decrease was attributed to the generation of new point defects in the previously-healed region, due to overlapping cascades. The healed-crack region comprised mainly hexagonal close-packed atoms, thus confirming effective healing following the initial cascade. A wealth of vacancies around the crack aided the formation of stacking-faults which limited defect-mobility during recombination. This was suggested the lower recombination-rate in cracked zirconium, as compared to that in crack-free zirconium. Collision-cascade induced healing required transformation of the original crack into a dislocation network which was enriched in vacancies. The accumulation of irradiation-induced point defects such as vacancies and interstitials led to dislocation-loop formation. The mechanical performance of irradiated cracked zirconium in uniaxial tensile simulations was compared with that of non-irradiated crack-free, irradiated crack-free and non-irradiated cracked material. The presence of pre-existing cracks and collision-cascade induced defects markedly reduced the yield stress. The yield stress of irradiated crack-free material decreased from the 12.15GPa of non-irradiated crack-free material to 8.09GPa following a single collision-cascade, illustrating the effect of residual Frenkel pairs on tensile properties. These pairs spoiled microstructural integrity by promoting stacking-fault formation and dislocation nucleation. In irradiated cracked material, the yield stress remained lower than that of non-irradiated material due to persistent vacancy clusters and a dearth of interstitials after healing. The initial crack acted, in effect, like a large vacancy-cluster with atoms on the periphery forming weak bonds. Under a tensile stress, these atoms migrated into interlayer spaces and increased the interlayer gap. Interstitials in these gaps could coalesce into interstitial clusters or dislocations and the presence of such vacancy clusters rendered the material more susceptible to deformation and reduced its strength.

About the Author

Dr. Fisher has wide knowledge and experience of the fields of engineering, metallurgy and solid-state physics, beginning with work at Rolls-Royce Aero Engines on turbine-blade research, related to the Concord supersonic passenger-aircraft project, which led to a BSc degree (1971) from the University of Wales. This was followed by theoretical and experimental work on the directional solidification of eutectic alloys having the ultimate aim of developing composite turbine blades. This work led to a doctoral degree (1978) from the Swiss Federal Institute of Technology (Lausanne). He then acted for many years as an editor of various academic journals, in particular *Defect and Diffusion Forum*. In recent years he has specialized in writing monographs which introduce readers to the most rapidly developing ideas in the fields of engineering, metallurgy and solid-state physics. He is co-author of the widely-cited student textbook, *Fundamentals of Solidification*. Google Scholar credits him with 8687 citations and a lifetime h-index of 14.

References

[1] Konkova V.A., Metallovedenie i Termicheskaya Obrabotka Metallov, 11, 1996, 30-32.

[2] Zhou G., Gao K., Wan F., Qiao L., Chu W., Progress in Natural Science, 11[3] 2001, 219-220.

[3] Bessone J.B., Corrosion Science, 48[12] 2006, 4243-4256. https://doi.org/10.1016/j.corsci.2006.03.013

[4] Zhao C., Zhu X., Langmuir, 41[10] 2025, 6592-6602. https://doi.org/10.1021/acs.langmuir.4c04419

[5] Yu L., Qiu-yang Z., Zhi-guo J., Zhi-peng Y., Hao-han S., Zhen-yu Z., Zhong-yu P., Journal of Materials Processing Technology, 339, 2025, 118803. https://doi.org/10.1016/j.jmatprotec.2025.118803

[6] Michalcová A., Marek I., Knaislová A., Sofer Z., Vojtěch D., Materials, 11[2] 2018, 199. https://doi.org/10.3390/ma11020199

[7] Suyitno, Eskin D.G., Katgerman L., Materials Science and Engineering A, 420[1-2] 2006, 1-7. https://doi.org/10.1016/j.msea.2005.12.037

[8] Li Y., Yao S., Chen H., Yang Y., Xu S., Zhang R., Engineering Fracture Mechanics, 288, 2023, 109346. https://doi.org/10.1016/j.engfracmech.2023.109346

[9] Gali O.A., Shafiei M., Hunter J.A., Riahi A.R., Materials Science and Engineering A, 618, 2014, 129-141. https://doi.org/10.1016/j.msea.2014.08.029

[10] Qbau N., Nam N.D., Ca N.X., Hien N.T., Journal of Manufacturing Processes, 50, 2020, 241-246. https://doi.org/10.1016/j.jmapro.2019.12.050

[11] Zhang D., Zhao X., Pan Y., Li H., Zhou L., Zhang J., Zhuang L., Metals, 10[2] 2020, 222. https://doi.org/10.3390/met10020222

[12] Taghiabadi R., Fayegh A., Pakbin A., Nazari M., Ghoncheh M.H., Transactions of Nonferrous Metals Society of China, 28[7] 2018, 1275-1286. https://doi.org/10.1016/S1003-6326(18)64783-1

[13] Hirano R., Ozaki S., Nagata Y., Matsushita A., Sakamoto T., Orio K., Okimura Y., Okane T., Faiz M.K., Yoshida M., Journal of Japan Institute of Light Metals, 73[9] 2023, 447-454. https://doi.org/10.2464/jilm.73.447a

[14] Lu Z., Guo C., Li P., Wang Z., Chang Y., Tang G., Jiang F., Journal of Alloys and Compounds, 708, 2017, 834-843. https://doi.org/10.1016/j.jallcom.2017.03.085

[15] Li P., Sui Y., Jiang Y., Yuan Y., Yang H., Yang J., Materials and Design, 246, 2024, 113325. https://doi.org/10.1016/j.matdes.2024.113325

[16] Ma K., Zhang X.X., Li X., Ma F., Chinese Journal of Nonferrous Metals, 24[2] 2014, 351-357.

[17] Deng Z., Xiao H., Yu C., Guo Y., Journal of Materials Research and Technology, 26, 2023, 935-947. https://doi.org/10.1016/j.jmrt.2023.07.243

[18] Kang H., Song K., Li L., Liu X., Jia Y., Wang G., Wang Y., Lan S., Lin X., Zhang L.C., Cao C., Journal of Materials Science and Technology, 179, 2024, 125-137. https://doi.org/10.1016/j.jmst.2023.08.048

[19] Wei D., Han J., Xie J., Hu H., Chen J., He Y., Journal of University of Science and Technology Beijing, 22[3] 2000, 245-248.

[20] Li S., Gao K.W., Qiao L.J., Zhou F.X., Chu W.Y., Computational Materials Science, 20[2] 2001, 143-150. https://doi.org/10.1016/S0927-0256(00)00130-0

[21] Gao K.W., Li S., Qiao L.J., Chu W.Y., Materials Science and Technology, 18[10] 2002, 1109-1114. https://doi.org/10.1179/026708302225006133

[22] Zhang Y., Han J., Liu J., Zhang L., Key Engineering Materials, 274-276[1] 2004, 817-822. https://doi.org/10.4028/www.scientific.net/KEM.274-276.817

[23] Zhang Y.J., Han J.T., Wei D.B., Journal of Iron and Steel Research, 16[2] 2004, 59-62.

[24] Li M., Chu W., Gao K., Su Y., Qiao L., Materials Letters, 58, 2004, 543-546. https://doi.org/10.1016/j.matlet.2003.06.016

[25] Boyko Y.I., Volosyuk M.A., Kononenko V.G., Functional Materials, 19[2] 2012, 245-250.

[26] Li J., Fang Q.H., Liu B., Liu Y., Liu Y.W., Wen P.H., Acta Materialia, 95, 2015, 291-301. https://doi.org/10.1016/j.actamat.2015.06.006

[27] Yuan X., Zhao Y., Kou S., Zhao Z., Li C.Y., Yuan Z., Rare Metal Materials and Engineering, 46[1] 2017, 35-38.

[28] Luo F., Zheng J., Chu G., Liu B., Zhang S., Li H., Chen L., Acta Chimica Sinica, 73[8] 2015, 808-814.

[29] Alcantar N.A., Park C., Pan J.M., Israelachvili J.N., Acta Materialia, 51[1] 2003, 31-47. https://doi.org/10.1016/S1359-6454(02)00225-2

[30] Lu Z.J., Liu L.Z., Xie H., Jin H.J., Acta Materialia, 292, 2025, 120997. https://doi.org/10.1016/j.actamat.2025.120997

[31] Gao K.W., Qiao L.J., Chu W.Y., Scripta Materialia, 44[7] 2001, 1055-1059. https://doi.org/10.1016/S1359-6462(01)00671-6

[32] Gao K., Qiao L., Chu W., Acta Metallurgica Sinica, 37[2] 2001, 118-120.

[33] Zhang H.L., Huang P.Z., Sun J., Gao H., Applied Physics Letters, 85[7] 2004, 1143-1145. https://doi.org/10.1063/1.1780592

[34] Nichols F.A., Mullins W.W., Transactions of the Metallurgical Society of AIME, 233, 1965, 1840.

[35] Wei D., Han J., Tieu A.K., Jiang Z., Scripta Materialia, 51[6] 2004, 583-587. https://doi.org/10.1016/j.scriptamat.2004.05.032

[36] Wei D., Jiang Z., Han J., Computational Materials Science, 69, 2013, 270-277. https://doi.org/10.1016/j.commatsci.2012.11.043

[37] Kan Y., Liu H., Zhang S.H., Zhang L.W., Cheng M., Journal of Manufacturing Science and Engineering, 135[5] 2013, 051003. https://doi.org/10.1115/1.4024764

[38] Xu J., Dong C., Li X., Journal of the Chinese Society of Corrosion and Protection, 24[4] 2004, 198-202.

[39] Kryzhevich D.S., Korchuganov A.V., Zolnikov K.P., Materials, 14[20] 2021, 6124. https://doi.org/10.3390/ma14206124

[40] Jiao Y., Zhou C., Liu J., Zhang X., Advances in Materials Science and Engineering, 2021, 2021, 8843885. https://doi.org/10.1155/2021/8843885

[41] Xin R.S., Kang J., Ma Q.X., Ren S., An H.L., Yao J.T., Pan J., Sun L., Metallurgical and Materials Transactions A, 49[10] 2018, 4906-4917. https://doi.org/10.1007/s11661-018-4814-x

[42] Li A., Chen X., Zhang C., Cui G., Zhao H., Yang C., Applied Surface Science, 442, 2018, 437-445. https://doi.org/10.1016/j.apsusc.2018.02.181

[43] Xu Q., Yuan X., Eckert J., Şopu D., Acta Materialia, 263, 2024, 119488. https://doi.org/10.1016/j.actamat.2023.119488

[44] Welker M., Rahmel A., Schütze M., Metallurgical Transactions A, 20[8] 1989, 1541-1551. https://doi.org/10.1007/BF02665510

[45] Chen H., He S., Chen J., Chen F., Zhang S., Zhang Y., Frontiers in Materials, 9,

2022, 1007502. https://doi.org/10.3389/fmats.2022.1007502

[46] Zhao Z., Perini M., Pellizzari M., Journal of Materials Science, 60[5] 2025, 2503-2523. https://doi.org/10.1007/s10853-024-10500-2

[47] Cai Q., Rey Rodriguez P., Carracelas Santos S., Castro G., Mendis C.L., Chang I.T.H., Assadi H., Materials Letters, 365, 2024, 136410. https://doi.org/10.1016/j.matlet.2024.136410

[48] Xu W., Yang C., Yu H., Jin X., Guo B., Shan D., Scientific Reports, 8[1] 2018, 6016. https://doi.org/10.1038/s41598-018-24354-7

[49] Xu W., Yang C., Yu H., Jin X., Yang G., Shan D., Guo B., Journal of Magnesium and Alloys, 9[5] 2021, 1768-1781. https://doi.org/10.1016/j.jma.2020.08.022

[50] Wei Z., Yang S., Gu H., Zhou L., Wang Z., Mu W., Wang F., Mao P., Engineering Failure Analysis, 177, 2025, 109678. https://doi.org/10.1016/j.engfailanal.2025.109678

[51] Guan B., Yang X., Tang J., Qin L., Xu M., Yan Y., Cheng Y., Le G., International Journal of Refractory Metals and Hard Materials, 112, 2023, 106123. https://doi.org/10.1016/j.ijrmhm.2023.106123

[52] Zheng X.G., Shi Y.N., Lu K., Materials Science and Engineering A, 561, 2013, 52-59. https://doi.org/10.1016/j.msea.2012.10.080

[53] Meraj M., Pal S., Applied Physics A, 123[2] 2017, 138. https://doi.org/10.1007/s00339-017-0760-5

[54] Liu S., Yang H., Chinese Physics B, 30[11] 2022, 116107. https://doi.org/10.1088/1674-1056/ac0780

[55] Chen P., Chesetti A., Demkowicz M.J., Journal of Nuclear Materials, 555, 2021, 153124. https://doi.org/10.1016/j.jnucmat.2021.153124

[56] Liu S.L., Wang C.Y., Yu T., Computational Materials Science, 110, 2015, 261-269. https://doi.org/10.1016/j.commatsci.2015.08.037

[57] Sukhov D.I., Petrushin N.V., Zaitsev D.V., Tikhonov M.M., Metallurgist, 63[3-4] 2019, 409-421. https://doi.org/10.1007/s11015-019-00837-4

[58] Alexandrov B.T., Hope A.T., Sowards J.W., Lippold J.C., McCracken S., Welding in the World, 55[3-4] 2011, 65-76. https://doi.org/10.1007/BF03321288

[59] Wheeling R.A., Lippold J.C., Materials Characterization, 115, 2016, 97-103. https://doi.org/10.1016/j.matchar.2016.03.006

[60] Wheeling R.A., Lippold J.C., Welding in the World, 61[2] 2017, 315-324. https://doi.org/10.1007/s40194-016-0411-z

[61] Wheeling R.A., Lippold J.C., Welding in the World, 64[1] 2020, 83-93. https://doi.org/10.1007/s40194-019-00806-0

[62] Vilanova M., Garciandia F., Sainz S., Jorge-Badiola D., Guraya T., San Sebastian M., Journal of Materials Processing Technology, 300, 2022, 117398. https://doi.org/10.1016/j.jmatprotec.2021.117398

[63] Hu X., Guo C., Huang Y., Xu Z., Shi Z., Zhou F., Li G., Zhou Y., Li Y., Li Z., Li Z., Lu H., Ding H., Dong H., Zhu Q., Acta Materialia, 267, 2024, 119731. https://doi.org/10.1016/j.actamat.2024.119731

[64] Wei D., Zhou W., Kong D., Tian Y., He J., Wang R., Huang W., Tan Q., Zhu G., Sun B., Journal of Materials Science and Technology, 210, 2025, 58-71. https://doi.org/10.1016/j.jmst.2024.05.042

[65] Wang L., Quan M., Tan Z., Liu M., Wang D., Yang X., Liu Y., Mao Y., Liang Z., Yang F. Journal of Materials Research and Technology, 31, 2024, 733-738. https://doi.org/10.1016/j.jmrt.2024.06.132

[66] Xi X., Lin D., Song X., Wei H., Yan M., Luo W., Chen B., Tan C., Dong Z., Minami F., Additive Manufacturing, 98, 2025, 104638. https://doi.org/10.1016/j.addma.2025.104638

[67] Zheng X.G., Shi Y.N., Lu K., Journal of the Electrochemical Society, 162[6] 2015, D222-D228. https://doi.org/10.1149/2.1021506jes

[68] Hsu K.T., Wang H.S., Chen H.G., Chen P.C., Metals, 6[10] 2016, 238. https://doi.org/10.3390/met6100238

[69] Gerstgrasser M., Cloots M., Stirnimann J., Wegener K., Journal of Materials Processing Technology, 289, 2021, 116948. https://doi.org/10.1016/j.jmatprotec.2020.116948

[70] Koo J., Kim J.E., Auyeskhan U., Park S., Jung I.D., Kim N., International Journal of Precision Engineering and Manufacturing, 26[3] 2025, 513-525.

[71] Xu G., Demkowicz M.J., Extreme Mechanics Letters, 8, 2016, 208-212. https://doi.org/10.1016/j.eml.2016.03.011

[72] Cronin H.M., Stoeva Z., Brown M., Shkunov M., Silva S.R.P., ACS Applied Materials and Interfaces, 10[25] 2018, 21398-21410. https://doi.org/10.1021/acsami.8b04157

[73] Xu D.E., Chow J., Mayer M., Jung J.P., Yoon J.H., Electronic Materials Letters, 11[6] 2015, 1078-1084. https://doi.org/10.1007/s13391-015-5201-z

[74] Zhang S., Choi H., Wang J., Liu Z., Han H.N., Hong S.T., International Journal of Fatigue, 198, 2025, 108967. https://doi.org/10.1016/j.ijfatigue.2025.108967

[75] Huang Z., Meng L., Lin P., Cao X., Zhang Z., Liu G., Journal of Alloys and Compounds, 938, 2023, 168604. https://doi.org/10.1016/j.jallcom.2022.168604

[76] Deng D., Yu T., Zhang L., Yang S., Zhang H., Journal of Mechanical Engineering, 53[18] 2017, 93-98. https://doi.org/10.3901/JME.2017.17.093

[77] Deng D., Yu T., Zhang L., Yang S., Zhang H., Journal of Mechanical Engineering, 53[20] 2017, 38-44. https://doi.org/10.3901/JME.2017.11.044

[78] Zu Q., Zhang H., Zha S., Liu S., Qi X., Zhao L., Chinese Journal of Theoretical and Applied Mechanics, 57[3] 2025, 712-719.

[79] Guo S., Lu J., Song Y., Xie L., Yu Y., Wang Z., Lu S., Engineering Failure Analysis, 159, 2024, 108120. https://doi.org/10.1016/j.engfailanal.2024.108120

[80] Vaidya R.U., Zurek A.K., Wolfenden A., Cantu M.W., Journal of Materials Engineering and Performance, 6[1] 1997, 46-50. https://doi.org/10.1007/s11665-997-0031-2

[81] Zhang X., Su C., Tian X., Chen M., Wan W., Yuan B., Materials Today Communications, 42, 2025, 111276. https://doi.org/10.1016/j.mtcomm.2024.111276

[82] Zhao Z., Wang G., Hou H., Zhang Y., Wang Y., Scientific Reports, 8[1] 2018, 14748. https://doi.org/10.1038/s41598-018-24089-5

[83] Wang H., Song G., Tang G., Journal of Alloys and Compounds, 681, 2016, 146-156. https://doi.org/10.1016/j.jallcom.2016.04.067

[84] Li S., Wang J., He J., Xue R., Wang R., Niu D., Chen R., Tang Y., Bai S., Intermetallics, 159, 2023, 107912. https://doi.org/10.1016/j.intermet.2023.107912

[85] Dunham T.E., Hehemann R.F., Metallurgical Transactions, 5[11] 1974, 2365-2373. https://doi.org/10.1007/BF02644018

[86] Chen J., Li K., Wang Y., Xing L., Yu C., Liu H., Ma J., Liu W., Shen Z., International Journal of Refractory Metals and Hard Materials, 87, 2020, 105135. https://doi.org/10.1016/j.ijrmhm.2019.105135

[87] Wang H., Qin C., Zhou Y., Mi X., Wang Y., Kang J., Pan R., Wu L., She J., Tan J., Tang A., Computational Materials Science, 214, 2022, 111688.

https://doi.org/10.1016/j.commatsci.2022.111688

[88] Li W., Zhao H., Zeng X., Yang X., Chi M., Gao Y., Nuclear Instruments and Methods in Physics Research B, 559, 2025, 165608. https://doi.org/10.1016/j.nimb.2024.165608